Production Management Systems

A CIM PERSPECTIVE

Production Management Systems

A CIM PERSPECTIVE

JIMMIE BROWNE

University College Galway
Galway, Ireland

JOHN HARHEN

Digital Equipment Corporation
Marlboro, MA., USA

JAMES SHIVNAN

Digital Equipment International
Galway, Ireland

ADDISON-WESLEY PUBLISHING COMPANY

Wokingham, England · Reading, Massachusetts · Menlo Park, California
New York · Don Mills, Ontario · Amsterdam · Bonn
Sydney · Singapore · Tokyo · Madrid · San Juan

To: Maeve, Lorcan and Shane,
Anna and Aoife Maire,
Ann.

Cover designed by John Gibbs and printed by The Riverside Printing Co. (Reading) Ltd.
Text design by Wendy Bann.
Typeset by Quorum Technical Services, Cheltenham, Glos.
Printed in Great Britain by T. J. Press (Padstow), Cornwall.

First printed 1988

British Library Cataloguing in Publication Data
Browne, Jimmie,
Production management systems.
1. Manufacture. Applications of computer systems.
I. Title II. Harhen, John
III. Shivnan, James
670.42′7
ISBN 0–201–17820–6

Library of Congress Cataloging in Publication Data
Browne, Jimmie.
Production management systems.

Bibliography: p.
Includes index.
1. Production management—Data processing.
2. Computer integrated manufacturing systems.
I. Harhen, John. II. Shivnan, James. III. Title.
TS155.6.B76 1988 658.5 88–16709
ISBN 0–201–17820–6

Foreword

It is easy to recognize the increasing degree of competition in global markets between rival manufacturing firms. Manufacturing management is faced with the difficult task of making wise investments in the various emerging technologies that are being offered as the competitive way forward for the manufacturing enterprise. To cope, manufacturing management must firstly educate itself as to what each wave of technology comprises. Secondly, management must attempt to see things in perspective, and position its response, so as to attain the strategic advantage of having the technological high-ground over its competitors, while at the same time maintaining competitiveness in the short term.

It is very opportune that this book comes along just as it does, and presents an integrative view of several of the key technologies that are driving manufacturing today. Manufacturing is on a journey from Class A MRP II, to just in time manufacturing, and the computer integrated enterprise, and beyond, to manufacturing excellence. It is absolutely clear that we will never achieve manufacturing excellence unless we continue to invest in developing the skills and deepening the knowledge of all our manufacturing people. This book serves an essential role in this developmental process, and it provides insights to manufacturing management as to what should be the way forward in applying these strategic technologies.

I have known the authors for several years now. Jimmie Browne has long been active internationally in Europe and beyond, in the CIM and production management domains, and John Harhen and James Shivnan have worked on challenging real problems in the CIM, AI and production management space. I note with satisfaction that, despite being geographically separated by great distance, Digital's internal computer network has facilitated their interaction across the Atlantic as this book developed. As a manufacturing manager who must deal with the issues that this book addresses, I am more than pleased to have been given the opportunity to write the Foreword. In conclusion, the book discusses in a pragmatic fashion the concepts essential for successfully undertaking the journey to manufacturing excellence.

Ed O'Connell
Computer Systems Manufacturing,
Digital Equipment International B.V.,
Galway, Ireland

Preface

Increasing emphasis is being placed on the manufacturing function in the competition between industrial firms. In order to compete successfully in global markets, firms must achieve excellence in managing their manufacturing operations. Computer Integrated Manufacturing (CIM) is seen as one of the key strategies that firms should adopt in their efforts to achieve manufacturing excellence. CIM involves the integrated application of computer technology to achieve the firm's business objectives. At the very heart of the CIM system lies the Production Management System (PMS), which regulates the pulse of the manufacturing firm through its decisions of what and when to buy and make. It comes as no surprise, therefore, that an effective production management system is essential for the firm in its efforts to achieve CIM and boost its competitiveness. This book is concerned with the various production management approaches available to firms engaged in discrete parts manufacturing.

The five-part structure of the book is illustrated in Figure I. In Part I we discuss the *new* competitive manufacturing environment in which firms find themselves and the emergence of CIM as a means to respond to this environment. This is the context in which future production management systems will operate. In industry, three distinct production management system strategies have emerged, which are seen to be competing. These strategies are Manufacturing Resource Planning (MRP II), Just in Time (JIT) and Optimized Production Technology (OPT). In the central sections of this book we take each of these strategies in turn, discuss its underlying philosophy and try to outline the basic techniques that it uses. Thus Parts II, III and IV deal with the MRP II, JIT and OPT approaches, respectively, and each may be read in isolation.

Part V seeks to offer a critical overview of the three approaches and to present some ideas for the design and operation of PMS systems in a CIM environment. In particular, we argue for an approach that puts equal emphasis on the social and technical subsystems in the design of production management systems. Furthermore we believe that a hybrid production management system, which draws on the best insights from JIT, MRP and

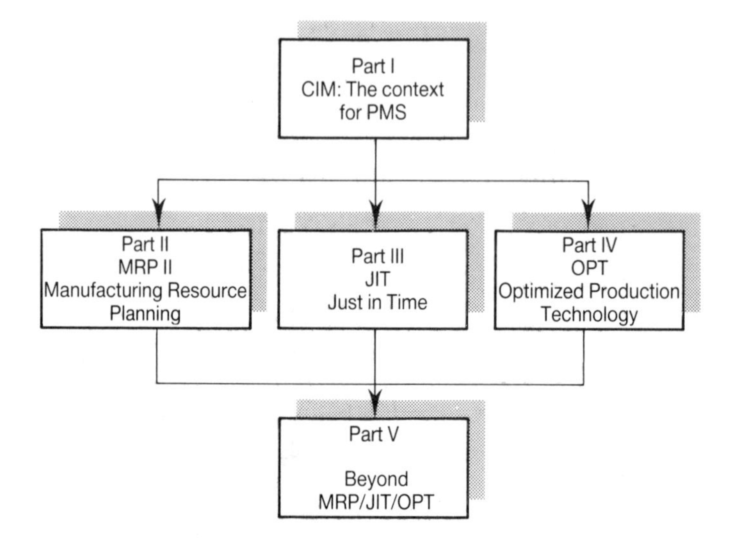

Figure I The structure of the book.

OPT, is likely to offer the best solution in the factory of the future. We present a preliminary sketch of what that hybrid system might look like.

The book is aimed primarily at three types of reader. Firstly, those professionals in industry, i.e. managers, engineers and production management personnel, who wish to explore the fundamentals of the various alternative and competing production management strategies for discrete parts manufacturing and who wish to develop ideas that can be used to good effect in their own firms. Secondly, students in industrial/manufacturing systems engineering and business schools, i.e. those involved in final year undergraduate and postgraduate industrial engineering and manufacturing engineering courses; also MBA students who are interested in the practice of production and operations management and manufacturing systems design. Finally, those taking specialized training for production management and manufacturing systems certification programs and those involved in industrial training courses in production and inventory management.

In writing this book, we have adopted an informal (i.e. non-mathematical) approach and have been guided constantly by the realization that, given the nature of production management, any proposed advances are only useful in so far as they find successful application in manufacturing industry. We believe that this book should help facilitate the transition from theory to practice in production and inventory management – at least that was our intention in writing it.

Acknowledgements

The journey towards writing this book involved many people over the last few years. We acknowledge our debt to those who helped us along the way and influenced the formation of our understanding of production management systems and their role within CIM systems.

We thank our colleagues at Digital Equipment Corporation, John McCahill, Manus Harley, Ed O'Connell, Dennis O'Connor and Bob Haynes, for the support that they have given us over that time. We thank Pat Galvin for his invaluable insights into the practice of production management. We particularly thank Richard Joyce, who made a very valuable contribution in the early stages of the formulation of this book.

We thank our colleagues within the Department of Industrial Engineering at University College Galway, in particular Dr Richard Gault, Mr John Roche and Professor M. E. J. O'Kelly for the many insights we have gained from them over the years. We are grateful to the Computer Services Department within UCG for their technical support in preparing the original manuscript.

We acknowledge the insights we have gained from many formal and informal discussions with colleagues from various European industries, universities, research institutes and the Commission of the European Community working within the ESPRIT programme of the European Economic Community. In particular those colleagues involved in COSIMA and other related ESPRIT CIM projects, who have offered us useful insights into the nature of production control problems in advanced manufacturing systems; also Professor Guissippe Menga from the Politecnico di Torino, Alfred Bauer from Digital Equipment Corporation in Munich, FRG, Dick Davies from Digital Equipment Corporation in Reading, UK, and Eric Gerelle from Digital Equipment Corporation in Geneva, Switzerland.

We also acknowledge the significant influences of our colleagues in IFIP Working Group 5.7, in particular Professors Peter Falster in Denmark, Asbjorn Rolstadas in Norway, John Burbidge in England, Guy Doumeingts in France and J.C. Worthmann in the Netherlands. We also thank those who have influenced our thinking over the years, including Professor John Davies and Dr Harry Jagdev of the University of Manchester Institute of Science and Technology, UK, Professor Andrew Kusiak in the University of Manitoba, Canada, Professor Kathryn Stecke in the University of Michigan, USA and Professor Keith Rathmill, UK.

We especially thank Professor Robert Graves from the University of Massachusetts for his very helpful criticism of the first draft of this book.

A special thanks goes to the researchers at the CIM Research Unit of University College Galway, particularly John Lenihan, Richard Bowden, Subhash Wadhwa and Jim Duggan, for their contribution to our understanding of CIM problems.

We thank Simon Plumtree, Allison King, Tim Pitts, Sheila Chatten and Margaret Conn from Addison-Wesley for their continuing patience and help as the manuscript developed.

We acknowledge the many authors whose work we have consulted in the preparation of this book and whom we have referenced in the manuscript. In particular we acknowledge Yasuhiro Monden for his book *Toyota Production Systems* from which we learned much about Just in Time, and Robert E. Fox whose articles were a clear presentation of the underlying concepts of Optimized Production Technology.

Last, but not least, we thank Maeve, Anna and Ann for suffering without complaint through this book's lengthy gestation period and without whose support we could not have continued.

Jimmie Browne, John Harhen, James Shivnan
January 1988.

Publisher's Acknowledgements

The publisher would like to thank the following for giving their permission to reprint their material:

Harvard Business Review for excerpts from (i) 'The Focused Factory' by Wickham Skinner (May–June 1974). Copyright © 1974 by The President and Fellows of Harvard College; all rights reserved. (ii) 'Plan for Economies of Scope' by Joel D. Goldhar and Mariann Jelinek (Nov.–Dec. 1983). Copyright © 1983 by The President and Fellows of Harvard College; all rights reserved. (iii) 'Manufacturing – the Missing Link in Corporate Strategy' by Wickham Skinner (May–June 1969). Copyright © 1969 by The President and Fellows of Harvard College; all rights reserved.

Elsevier Science Publishers for excerpts from (i) Primrose, P. and Leonard R. (1986) 'Conditions Under Which Flexible Manufacturing is Financially Viable' in *Flexible Manufacturing Systems: Methods and Studies* edited by A. Kusiak. (ii) Rosenthal, S. and Ward, P. (1986) 'Key Managerial Roles in Controlling Progress Towards CIM' in *Manufacturing Research: Organizational and Institutional Issues* edited by A. Gerstenfeld, H. Bullinger and H. Warnecke. (iii) Burbidge, J. (1985a) 'Automated Production Control' in *Modelling Production Management Systems* edited by P. Falster and R. Mazumber. (iv) for excerpts from and Figures 2.1 and 2.2 and Table 2.1 taken from figures in Harhen, J. and Browne, J. (1984) 'Production Activity Control: A Key Node in CIM' in *Strategies for Design and Economic Analysis of Computer Supported Production Management Systems* edited by H. Hubner. (v) Shivnan, J., Joyce, R. and Browne, J. (1987) 'Production and Inventory Management Techniques – A Systems Perspective' in *Modern Production Management Systems* edited by A. Kusiak.

Pergamon Journals Ltd for an excerpt from Spur, G. (1984) 'Growth, Crisis and the Factory of the Future', *Robotics and Computer Integrated Manufacturing*, **1** (1), 21–37.

Computer Aided Manufacturing International, Inc. for an excerpt from CAM-I Factory Management Project, PR-82-ASPP-01.6.

Taylor & Francis for an excerpt from Meredith, J. and Suresh, N. (1986) 'Justification Techniques for Advanced Manufacturing Technologies', *International Journal of Production Research*, **24** (5), 1043–57.

American Production and Inventory Control Society, Inc. for excerpts from *Production and Inventory Management*: (i) Vollman, T. E. (1986) 'OPT as an Enhancement to MRP II', **27** (2), 38–46; (ii) Swann, D. (1986) 'Using MRP for Optimized Schedules (Emulating OPT)', **27** (2), 30–37; (iii) Galvin, P. (1986) 'Visions and Realities: MRP as System', **27** (3); (iv) Hinds, S. (1982) 'The Spirit of Materials Requirements Planning', **23** (4), 35–50; (v) Latham, D. (1981) 'Are You Among MRP's Walking Wounded?', **22** (3), 33–41; (vi) Safizadeh, M. and Raafat, F. (1986) 'Formal/Informal Systems and MRP Implementation', **27** (1); (vii) Benson, P., Hill, A. and Hoffman, T. (1982) 'Manufacturing Systems of the Future: A Delphi Study', **23** (3), 87–106; (viii) Mather, H. (1985), 'Dynamic Lot Sizing for MRP: Help or Hindrance', **26** (2); (ix) St. John, R. (1984) 'The Evils of Lot Sizing in MRP', **25** (4); (x) DeBodt, M., van Wassenhove, L. (1983) 'Lot Sizes and Safety Stocks in MRP', **24** (1). Also for an excerpt from Wallace, T. (1980) ed. *APICS Dictionary* 4th edn.

IFS (Publications) for Figure 2.4 based on a figure from *The FMS Magazine*, April 1984; IFS (Conferences) for Figure 16.4 based on a figure from the 4th Automan Conference Proceedings.

Dow-Jones Irwin for an excerpt from Hall, R. W. (1983) *Zero Inventories*.

Butterworth Scientific Ltd for an excerpt from Gallagher, C. C. and Knight, W. A. (1973) *Group Technology*.

Society of Manufacturing Engineers for an excerpt from Laszcz, J. Z. (1985) 'Product Design for Robotic and Automatic Assembly', in *Robotic Assembly* edited by K. Rathmill, IFS Publications Ltd. Copyright © 1984, from the Robots 8 Conference Proceedings.

Institute of Industrial Engineers for (i) an excerpt from and (ii) Figures 15.1, 15.2 and 15.3 based on figures taken from Jacobs, F. R. (1984) 'OPT Uncovered: Many Productions Planning and Scheduling Concepts Can Be Applied With or Without the Software', *Industrial Engineering*, **16** (10). Copyright Institute of Industrial Engineers, 25 Technology Park/Atlanta, Norcross, GA 30092. (iii) for Figures 12.9, 13.4 and 13.25 based on figures taken from *Toyota Production Systems*, Monden, Y. Copyright 1983 Institute of Industrial Engineers.

Penton Publishing for an excerpt from (1981) 'Implementing CIM', *American Machinist*, 152–74.

W. H. Freeman and Company for an excerpt from Gunn, T. (1982). 'The Mechanization of Design and Manufacturing', *Scientific American*, **247** (3), 87–110.

Institute of Electrical and Electronics Engineers, Inc. for an excerpt from Cortes-Comerer, N. (1986) 'JIT is Made to Order', *IEEE Spectrum*, September, 57–62.

Auerbach Publishers, Inc. for an excerpt from Gold, B. (1986) 'CIM Dictates Change in Management Practice', *CIM Review*, **2** (3), 3–6.

Wright Publications and R. E. Fox for (i) Figures 14.2, 14.3 and 14.6 taken from figures in Fox, R. E. (1982) 'OPT – An Answer for America' Part II, *Inventories and Production Magazine*, Nov.–Dec. 1982. (ii) for an excerpt from 'MRP, Kanban or OPT. What's Best?', *Inventories and Production Magazine*, July–Aug. 1982.

About the Authors

Jimmie Browne is a lecturer in production engineering and Director of the CIM Research Unit at University College Galway in Ireland. He received his Ph.D. degree from the University of Manchester in 1980 for work on simulation of manufacturing systems. His research for the past 12 years has focused on the role of production management systems in CIM. He has authored and co-authored many publications in the areas of production management, flexible assembly systems, computer integrated manufacturing, simulation and artificial intelligence (AI) applications in manufacturing. He is a member of CIM Europe and also of IFIP Working Group 5.7 on Production Planning and Control.

James Shivnan is an engineer working with the Advanced Manufacturing Technology Group of Digital Equipment International at Galway, Ireland. He is currently working on production activity control systems for a CIM environment, simulation and AI approaches to manufacturing process design. He has worked on the comparison of alternate production management paradigms and their effectiveness for production activity control. James has publications in the areas of production management systems, simulation and AI applications. James holds a Master of Engineering Science degree in Industrial Engineering from University College, Galway.

John Harhen is a principal engineer working with the Intelligent Systems Technology Group of Digital Equipment Corporation in Marlboro, Massachusetts. He is currently applying knowledge based technology to strategic decision processes in the manufacturing function at Digital. His previous activities during his eight years with Digital include working on the implementation and education for MRP II at Digital's Clonmel plant in Ireland – an implementation that is now rated as *Class A*. John received a Master of Engineering Science degree in Industrial Engineering from University College, Galway, and is currently pursuing a Ph.D. from the University of Massachusetts in Amherst. John is certified at fellowship level in production and inventory management (CFPIM) by the American Production and Inventory Control Society (APICS).

Contents

Trademark Notice

PART I

CIM: the context for production management

Overview

Production management takes place in a context. This context comprises both the nature of discrete parts manufacturing, as well as the technological framework within which the Production Management System (PMS) finds itself, namely Computer Integrated Manufacturing (CIM). Part I of this book is devoted to discussing this context. It is structured as follows.

Chapter 1 discusses the emerging competitive pressures that are acting upon the discrete parts manufacturing firm. Various organizations of the manufacturing process are possible in this mode of manufacturing. These are jobbing shop production, batch production and mass production. In discussing the features of, and differences between, these types of process organization, the changes brought about by the new emphasis on competitiveness in manufacturing will be highlighted.

In Chapter 2, the development of automation in discrete parts manufacturing is discussed. The various stages on the road to CIM are identified. Significant islands of automation are described. Particular attention is devoted to a discussion of Flexible Manufacturing Systems (FMS). An examination of the history of the development of manufacturing automation should provide an understanding of the current emphasis on achieving computer integrated manufacturing. All of this serves to set the stage for a discussion of CIM.

In Chapter 3, CIM is examined from various viewpoints, such as the engineering perspective, and also the production management perspective. CIM is described in the narrow sense as **four walls** CIM, as well as in the broader view as the integration of factory CIM with the rest of the business. Both the single plant and multiplant cases of computer integrated business (CIB) are discussed.

Finally, Chapter 4 discusses the role of production management within the CIM system. The structure of the PMS is examined and several different levels are identified. Production management is identified as a central function within CIM. Some attention is then devoted to the relationship between the production management system and the nature of the manufacturing process.

The execution level, Production Activity Control (PAC), is seen as the gateway between the higher level planning modules within the PMS and the process technology of the shop floor; PAC, therefore, is the key to the achievement of CIM.

Because the production management system is at the heart of CIM, it is therefore very important to the competitive operation of the firm. The choice of an appropriate production management strategy can determine the success or failure of the firm. For this reason, three important alternative strategies for production management will be examined in detail. These are Manufacturing Resource Planning (MRP II), Just in Time (JIT) and Optimized Production Technology (OPT). This comprises the material for the next three parts of this book. Finally, in Part V, the assumptions, approaches and strategies implicit in the use of these techniques will be compared and contrasted.

CHAPTER ONE

The new manufacturing environment

1.1 Introduction

Writing in the *Harvard Business Review*, Skinner (1969) pointed to the unusual position of manufacturing within the business environment at that time. He wrote:

> 'A company's manufacturing function typically is either a competitive weapon or a corporate millstone. It is seldom neutral . . . Few top managers are aware that what appear to be routine manufacturing decisions frequently come to limit the corporation's strategic options, binding it with facilities, equipment, personnel and basic controls and policies to a non-competitive posture which may take years to turn around'.

Recent developments, particularly the success of Japan in world markets and the feeling that this success has been derived, to a significant degree, from superior manufacturing systems, have changed people's perception of the role and importance of the manufacturing function in the industrial firm. Business managers and business authors, such as Hayes and Wheelright (1984), now regard manufacturing as a competitive weapon in the marketplace and recommend that each company include in their business plans specific goals in the area of achieving manufacturing excellence. Manufacturing is now an equal partner at the corporate boardroom. The focus in manufacturing is competitiveness and the emphasis is on CIM and the factory of the future as the means of achieving it.

Skinner, writing in 1985, acknowledged the new situation:

> 'After years of neglect, top management's attention has been captured . . . the action in manufacturing has been extraordinary in these last five years'.

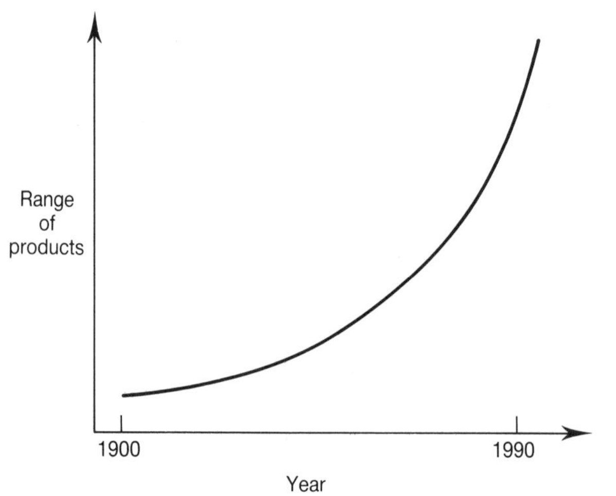

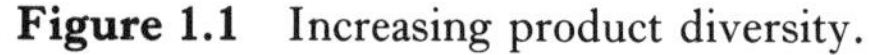

Figure 1.1 Increasing product diversity.

1.2 The new environment for manufacturing

Manufacturing is therefore seen as the new competitive weapon and, as a result, manufacturing firms find themselves in a totally changed environment. This change is not confined to any one industry and evidence of it can be seen in such varied industries as the automotive industry, consumer goods, electronics and white goods. Management faced with rapid changes must devise new strategies to deal with the competitive nature of this new environment. The old strategy of mass production, derived from notions of economies of scale, is no longer seen as valid and is being discarded in favour of a strategy which facilitates **flexibility**, reduced **design cycle time**, reduced **time to market** for new products and reduced **order cycle time** to the customer for existing products.

Some important characteristics of this new environment are:

- Increased product diversity.
- Greatly reduced product life cycles.
- Changing cost patterns.
- Great difficulty in estimating the costs and benefits of CIM technology.

Each of these factors will now briefly be discussed in turn.

Increased product diversity

The market is no longer satisfied with a mass produced uniform product. Manufacturing firms must now compete by offering variety. The age of the **personalized** consumer product seems to be rapidly approaching. This represents a much changed situation compared with earlier this century (see Figure 1.1).

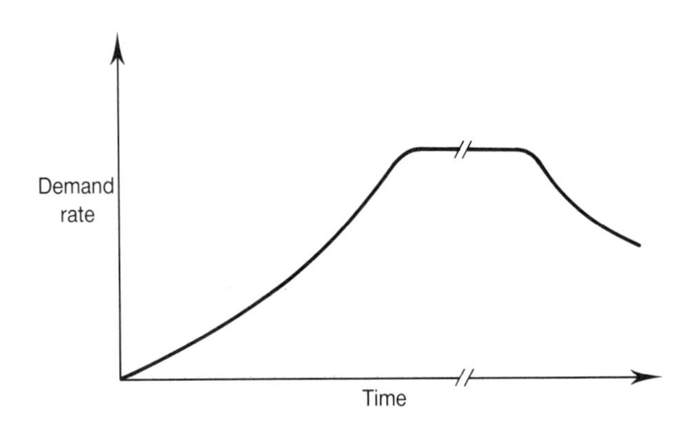

Figure 1.2 Traditional product life cycle.

The explosion of product variety is particularly evident in the automobile and computer industries. However, product variety dramatically increases the complexity of the tasks of process design and production management. For example, frequent process changeover can be a significant burden. Offering increased choice at reduced cost thus poses significant challenges for the manufacturing firm that is attempting to achieve or maintain competitiveness.

Greatly reduced product life cycles

The life cycle of a product falls naturally into several phases. In simple terms these are the design phase, the manufacturing phase and the end-of-life phase. The complete product life cycle, in simplified form, is represented by Figure 1.2. Demand is light and grows very slowly in the initial periods during which the manufacturer can establish the product design and production method. The second stage is one of a mature product enjoying high stable demand. The third and final stage sees the gradual decline in demand for the product. The costs incurred during the early part of the design cycle include the design costs and the costs associated with developing and installing the production process.

In the old way of doing business, the design cycle and the manufacturing cycle were separated and occurred *sequentially*. A product design was *proven* before it entered production. There was thus a significant amount of time available for production methods to be established prior to volume production of the product. The situation is illustrated in Figure 1.3. Once the product was established in the marketplace the manufacturer could look forward to a relatively high demand for a number of years before the product became obsolete. The many years of high, stable demand enabled the costs incurred in the early stage of product and process development to be recovered.

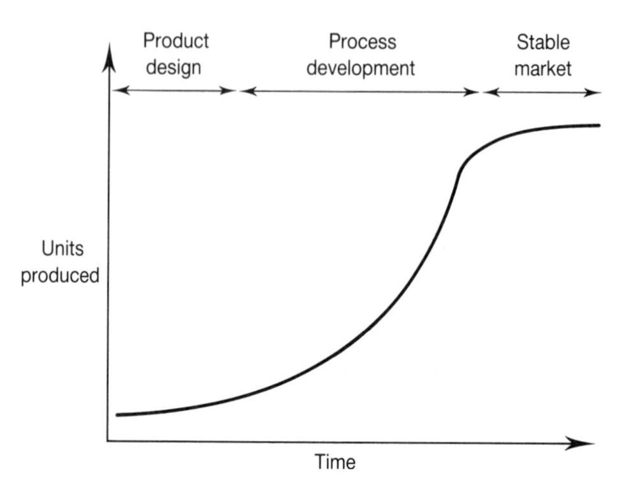

Figure 1.3 Development of market.

The difficulty in today's manufacturing environment is that manufacturing can no longer look forward to many years of stable, high demand. This is because product redesign is continuously happening and a product's useful life in the marketplace is constantly under attack from improved versions incorporating the latest design features. Moreover, due to the pressures of competition, firms must strive to get their products to the marketplace in ever decreasing times. All of this means that manufacturing must put processes in place which are sufficiently flexible to accommodate new product designs rapidly, without incurring large process introduction costs. Otherwise the costs incurred in the product design and process development phase will be too large to be recovered over the much shorter peak demand phase of the product life cycle. The cost distribution over this shortened life cycle is illustrated in Figure 1.4.

Because of the compression of product life cycles, manufacturing firms can no longer expend huge resources on developing a dedicated production capability, since the product design is likely to change before that production facility has been paid for. The concept of **economies of scale** has been replaced by the notion of **economies of scope**. Goldhar (1983) explains that economies of scope exist

> '. . . when the same equipment can produce the multiple products more cheaply in combination than separately. A computer controlled machine does not care whether it works in succession on a dozen different product designs – within, of course, a family of design limits. Changeover times (and therefore costs) are negligible, since the task of machine set-up involves little more than reading a computer program'.

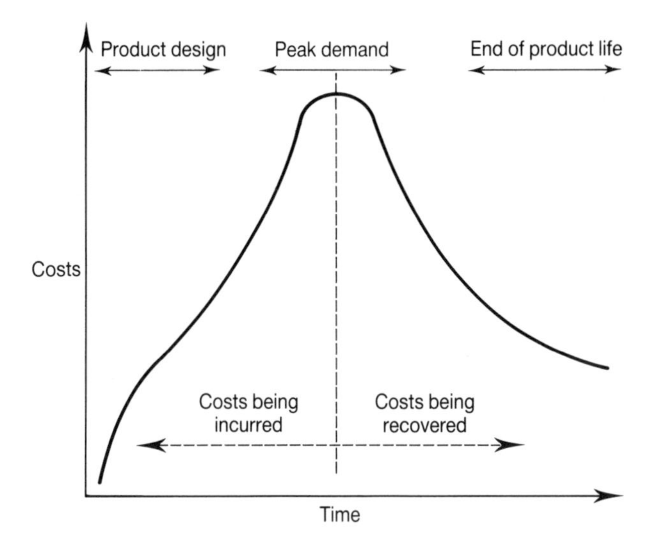

Figure 1.4 Distribution of cost over product life cycle.

Moreover, this has implications for investment appraisal procedures since companies must now put in place flexible production facilities which will be used not only on existing product designs, but also on future redesigns of these products. In summary, the combination of greater product diversity, shortened product life cycles and pressures for faster time to market have given rise to the need for economies of scope.

Changing cost structure

Manufacturing costs have traditionally fallen under three headings – material cost, labour cost and overhead. Furthermore, labour hours were used in many industries as a base for *recovering* overhead. Bonsack (1986) points out that 'current overhead accounting most often uses the **full absorption** method which assigns all factory overhead to units of production based on the labour costs incurred'.

What happens when direct labour costs fall to a very small fraction of total cost – to below 5%, as is the case in very many factories today? A manager in the electronics industry made the following observation, as quoted by Gould (1985): 'Direct labour used to account for almost half of production costs; in today's electronics it accounts for 5 to 8%. For example, IBM's direct labour cost is about 4%; at Apple's new Macintosh plant it is 1%'. In the factory of the future when almost unmanned manufacturing will be the norm, it seems obvious that direct labour cost can no longer be the foundation of a standard costing system. Bonsack (1986) speculates that it is likely that 'the direct labour cost category will eventually be eliminated entirely in many automated work cells by the incorporation of all costs except material in overhead'. Innovative costing methods will undoubtedly have to be developed that allocate this overhead cost against meaningful measures of output.

It can be argued that since CIM systems will incorporate sophisticated data capture technology at each production stage, this will consequently facilitate the association of cost with individual items – whether components, sub-assemblies or finished products – at the point of time at which the cost is incurred. This, however, ignores the issue of how to value a machine hour in an environment of rapid technological and product change, where there is great difficulty in predicting the useful life of the machine and, equally important, the life cycle of the products that are processed on it. All of this leads to the next issue.

The difficulty of estimating the costs and benefits of CIM

It is frequently very difficult to justify investment in CIM technology using traditional investment appraisal methods. This problem can be understood in the context of the economies of scope ideas proposed by Baumol and Braunstein (1977) and Goldhar and Jelinek (1985) and discussed above. Gold (1986) claims that 'field studies of several industries suggest that the fundamental reason that so many companies have failed to exploit CIM's potential is that management has generally assumed that the new technologies are no different than the traditional equipment acquired to improve production efficiency'. He goes on to say that when a company appraises CIM acquisition proposals, it often makes some false assumptions including:

- 'The prospective equipment will affect only a narrow range of production activities.
- The capabilities of the equipment are known and will not change significantly (except for gradual decline) after installation.
- The acquisition's contribution to the effectiveness of operations and cost reductions can be estimated with reasonable accuracy . . .'

These assumptions are clearly dubious for CIM technology. For example, CIM equipment is extremely flexible and thus the limits of its capability and its application are difficult to define. Meredith and Suresh (1986) suggest using differing justification strategies corresponding to the different levels of automation under consideration. Economic justification procedures are appropriate when a company is involved in straightforward replacement of old equipment, such as the purchase of a Numerically Controlled (NC) lathe to replace an existing conventional lathe. 'However, with systems approaching full integration, clear competitive advantages and major increments towards the firm's business objectives are usually being attained. Here strategic approaches are needed that take these benefits into consideration, although tactical and economic benefits may be accruing as well'.

Primrose and Leonard (1986), talking in terms of Flexible Manufacturing Systems (FMS), which can be seen as **mini-CIM** systems, point out that:

> 'Because of the company wide implications of introducing FMS, many authors have suggested that the benefits are considerable but intangible, with the word *intangible* being regarded as synonymous with *unquantifiable* . . . Because the inclusion of the intangibles significantly influences the viability of a proposed FMS . . . (it becomes) necessary to develop techniques to enable all of the potential benefits of such a system to be both quantified and included in an evaluation'.

We have now identified a number of important characteristics of this new emerging competitive environment. This environment has significant implications for the manner in which industrial firms organize their manufacturing processes. Traditional styles of manufacturing organization and their effect on manufacturing competition will now be discussed.

1.3 Manufacturing process organization for discrete manufacturing

Manufacturing systems analysts, such as Wild (1971), have identified two basic categories of industrial plant, namely continuous process industries and discrete parts manufacturing. Continuous process industries involve the continuous production of product, often using chemical rather than physical or mechanical means (e.g. the production of fertilizers or sugar). Discrete parts production involves the production of individual items and is further subdivided into mass, batch and jobbing shop production, as illustrated in Figure 1.5. Our focus in this book is on discrete parts manufacturing.

1.3.1 Jobbing shop production

The main characteristic of jobbing shop production is very low volume production runs of many different products. These products have a very low level of standardization in that there are few, if any, common components. To produce the different products, the manufacturing firm requires a highly flexible production capability. This implies flexible equipment capable of performing many different tasks, as well as a highly skilled work force. Jobbing shops normally operate a **make to order** or **engineer to order** inventory policy. A typical example of the jobbing shop is a subcontract machine shop.

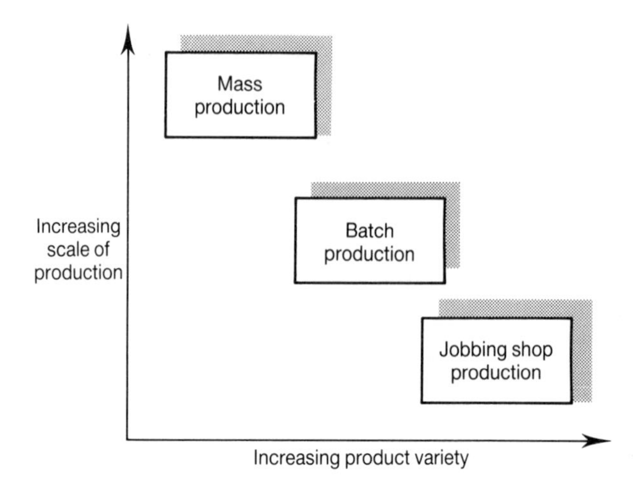

Figure 1.5 Classification of discrete production.

1.3.2 Batch production

Batch production's main characteristics are medium volume production runs of a medium range of products. Batch production is defined as the production of a product in small batches or lots by a series of operations, each operation typically being carried out on the whole batch before any subsequent operation is started.

The production system must be reasonably flexible and use general purpose equipment in order to accommodate varying customer requirements and fluctuations in demand. Batch production can be seen as a situation which lies between the extremes of the pure jobbing shop and pure mass production, where the quantity required is insufficient to justify mass production. Batching, however, offers economies in terms of amortizing set-up cost. Because of the large variety of jobs involved, batch production has much of the complexity of the jobbing shop. A typical example of batch production is the manufacture and assembly of machine tools.

Batch production represents a sizeable element within the total manufacturing base of developed economies. Gerwin (1982) cites statistics which claim that batch production represents more than 35% of the US manufacturing base and constitutes 36% of manufacturing's share of the GNP.

Furthermore, there is some evidence that companies that previously have used mass production methods are being forced by the pressures of the market to adopt more flexible batch production oriented systems, i.e. systems capable of dealing with relatively small quantities of a variety of products. This is particularly true of the automotive industry and manufacturers of consumer goods.

1.3.3 Mass production

The major characteristics of mass production are large volume production runs of relatively few products. All products are highly standardized. Typically, demand is stable for the products and the product design changes very little over the short to medium term.

The production facilities consist of highly specialized, dedicated machines. Although these machines are extremely expensive, the cost is amortized over very long production runs. The term **hard automation** or **Detroit style automation** was coined to describe the type of automation associated with mass production. It is *hard* in the sense that the automation is dedicated and very inflexible. The classic example of mass production used to be automotive manufacture and assembly, hence the term Detroit style automation. Nowadays, for reasons which will briefly be explored later, the automobile industry is no longer solely bound to mass production strategies.

1.4 Comparing the types of manufacturing process organization

The three categories of mass, batch and jobbing shop production have been discussed as if they represented a clear picture of the categories of discrete parts manufacturing, into which all discrete parts manufacturing systems could be neatly slotted. The reality of manufacturing, of course, is somewhat more complicated. Manufacturing exists on a continuum between two extremes – jobbing shops and pure mass production – and the majority of discrete parts manufacturing facilities lies somewhere along that continuum, not fitting into a well defined category. In many cases, differing and hybrid manufacturing process organizations can be found in the same production facilities. Nonetheless, the important characteristics of the three types of discrete parts manufacturing are given in summarized form in Figure 1.6.

The extremes of mass production versus jobbing shop can also be explored in terms of the difference between manufacturing process organization based on product flow and manufacturing process organization based on the commonality of process equipment, as discussed by Rolstadas (1986). Traditionally, the process layout, which involves organizing the plant in terms of individual processes or operations, has been associated with batch production and jobbing shops, and results in great flexibility but long throughput times. The product oriented approach, which involves laying out the plant to correspond to the flow of products through the processes they require, has traditionally been associated with mass production systems and has, in recent years, through the development of group technology principles, begun to be applied to batch production systems. Product flow layouts are somewhat less flexible than the process oriented approach, but result in

	Mass production	Batch production	Jobbing shop production
Production volumes	High	Medium	Low
Labour skills	Low	Medium	High
Specialized equipment and tooling	High	Medium	Low

Figure 1.6 Characteristics of types of process organizations.

greatly reduced throughput times. These issues will be considered in more detail in Part III, when JIT (Just in Time) approaches to manufacturing problems will be discussed.

Because of new competitive pressures, product diversity is now the norm and manufacturing companies that may previously have used a mass production strategy are now having to move into a more batch oriented environment. This is to achieve more flexibility in product introduction and production scheduling. The automotive industry, which will now be considered briefly, illustrates this trend very well.

1.4.1 Changing trends in the automotive industry

When people think of mass production, the manufacture of cars in the early part of this century often springs to mind. The attitude of car manufacturers at that time was epitomized by Henry Ford. His byword that 'the customer could have any colour car he liked as long as it was black' offers an insight as to why the manufacture of automobiles was oriented to mass production. The automobile companies standardized their product to facilitate the manufacturing process. This philosophy allowed the manufacturers to use dedicated, specialized machines, with a relatively low skilled work force. Based on economies of scale, the car manufacturers could produce highly standardized cars in large volumes, at a price that potential customers could afford to pay. The classical example of mass production automation was, and remains, the transfer line. This particular example of hard automation was, and is still, widely used in the manufacture and fabrication of components and sub-assemblies in the automotive industry.

However, over the last 20 years, there have been many changes. Customers, perhaps encouraged by sophisticated marketing and sales campaigns, are demanding greater choice when buying cars, not only as regards the colour but also the degree of interior luxury, the number of doors (whether two door, four door or hatch-back), the engine size, manual or automatic gearbox, etc. This has resulted in each manufacturer being forced to offer a range of cars with a variety of models for each segment in the range. Also each model must be upgraded each year and major model changes are very frequent.

For these reasons, automobile production has shifted away from a mass production activity and is rapidly assuming the characteristics of batch production. Manufacturers can no longer standardize on a particular design and tool up their plants in anticipation of long product life cycles. The strategy of investing heavily in specialized (in a product model sense) equipment in anticipation of recouping this investment over a long production run is no longer seen as appropriate. The reader interested in a discussion on the consequences of programmable automation for the automotive industry is referred to Gerwin and Tarondeau (1986).

1.5 Conclusion

We have seen that competing through manufacturing is a new and essential feature of industrial competition. We have identified what we consider to be the important characteristics of the new environment in manufacturing, namely increased product diversity, significantly reduced product life cycles, changing cost structures and the difficulties of understanding the costs and benefits of the new technology. This is having profound implications for the way manufacturing processes are organized. In particular, there is a trend away from pure mass production organization towards batch oriented systems in order to provide a greater diversity of products to an increasingly more demanding marketplace. The issue of manufacturing process organization will be briefly discussed again in Chapter 4, when the consequences of the nature of process organization for production management will be examined.

CHAPTER TWO

Development of automation in manufacturing

2.1 Introduction

Computer Integrated Manufacturing (CIM) can be viewed as the culmination of a long and ongoing effort in the application of automation in the manufacture of discrete parts. Table 2.1 presents a view of this evolutionary process as comprising four stages – mechanization, point automation, islands of automation and computer integrated manufacturing (Harhen and Browne 1984). This chapter briefly describes the history of the automation effort and, in so doing, sets the stage for the discussion of computer integrated manufacturing in Chapter 3. Readers interested in a more detailed statement of the history of the development of mechanization and automation are referred to Hitomi (1979).

2.2 Mechanization

The search for better ways to manufacture components was always the main driving force behind automation. The industrial revolution was largely concerned with the replacement of human labour by machines. This replacement of human physical labour is the primary characteristic of the mechanization phase. In the early decades of this century, F.W. Taylor and others introduced many new techniques to help standardize the operations and work methods in manufacturing. Taylor is well known as the father of scientific management (and industrial engineering). His approach was systematically to divide the manufacturing operation into smaller and smaller elements and then to concentrate on improving each element in turn. Such an approach facilitates the mechanization and later automation of specific

Table 2.1 Stages of manufacturing automation.

Stage	Features	Examples	Date
Mechanization	Replacement of human labour by machine	Lathe Power Conveyors	1775
Point automation	Replacement of human control of machine by automatic control	NC/CNC MRP	1960
Islands of automation	Integration of point automation within its local environment to manage part of the manufacturing process	MRP II FMS CAD/CAM	1970
Computer integrated manufacturing	The integrated application of computer based automation and decision support systems to manage the total operation of the manufacturing system	The automated and the automatic factory	1990?

operations. However, this reductionist approach is not adequate for the complexity and competitive nature of today's manufacturing.

Culture may have played a significant part in the evolution of manufacturing automation by constraining the approach of engineers and managers to various problems and process developments. This is certainly an important consideration in the discussion of just in time as a production management approach. It is also a consideration when discussing approaches to the automation of manufacturing. For this reason, it is useful to make a small digression to ponder some differences between Eastern and Western thinking.

Western society has tended to adopt the world view of scientific method, which is reductionist, quantitative and analytic in nature. The major tenet of this approach is that the whole can be reduced to its constituent parts and each examined on its own. In this manner, it is assumed that the system itself is also understood. Such thinking is evident in the work of Adam Smith, who laid the basis for the division of labour at the beginning of the industrial revolution, and more particularly so in the case of F.W. Taylor and his approach to management at the beginning of this century. If one accepts this argument, then the approach of Western manufacturing systems experts can be seen as the focused examination of well defined areas, without giving due consideration to the overall system. For example, many of the efforts of quantitative Operations Research (OR) have not had significant impact on the practice of production scheduling (see King 1976), despite the fact that great energy has been expended over the last 20 years, since those initial influential formulations of scheduling problems in the 1960s by researchers such as Conway *et al.* (1967). Moreover, writers such as Burbidge (1986) have reiterated this theme by taking issue with the overspecialized nature of manufacturing personnel.

In contrast, Eastern society seems more often to adopt a **systems perspective** of the world. This world view holds that the whole is greater than the sum of its parts and so recognizes the importance of interaction between the constituent sub-systems. In such a **holistic** approach, each sub-system is seen as having a certain autonomy, while still operating within the overall goals of the system. The most important aspect is that no sub-system proceeds with an action which is detrimental to other sub-systems. This style of thinking is exhibited frequently in the Just in Time (JIT) approach to production management in Japan. For example, Shingo (1981) writing on Kanban at Toyota, declares, 'the following is considered quite important: (to) acknowledge the conception of **Toyota production system**, its techniques and besides the systematic relationship between each technique'.

The Western approach is perhaps best exemplified by so-called **mechanistic** work organization and work structures, where individual operators tend to be assigned to a few very specialized repetitive tasks in a hierarchical supervisory environment. The alternative approach, the so-called **organic** work organization, is characterized by multiskilled operators working in relatively autonomous work groups and under a less rigid control and supervisory organization. In the authors' experience, the latter approach seems more appropriate, particularly in modern manufacturing systems. The reader interested in more detailed discussion of these issues is referred to Nanda (1986), Bullinger and Ammer (1986) and Cross (1984). A parting comment is that since the CIM problem is primarily about integration, it follows that a holistic approach using organic work structures may well be necessary to attain an effective solution.

2.3 Point automation

With the introduction of control technology to the factory in the 1950s and 1960s, the human control of some machines was replaced by numerical or computer control. For example, conventional lathes could be replaced by Numerical Control (NC) and later by Computer Numerical Control (CNC) lathes.

Numerical control is a form of programmable automation where the operation of a machine is controlled by numbers or symbols. A collection of numbers forms a program that drives a machine to produce a part. When a new part is to be produced, a new set of program steps is used. This ability to change the program, rather than the machine, is the source of the greater flexibility that is characteristic of computer based or **soft** automation. The term *soft* refers to the fact that the device is under program or software control.

There is thus a clear contrast between hard automation, which is very inflexible, and the soft automation, which is more flexible. Hard automation involves a situation where the configuration of the equipment determines the feasible set of operations. If the range of products change or new operations are required, hard automation is unable to cope without extensive reconfiguration. Early approaches to automation were typically of the hard variety, the archetypical example being the transfer line (Groover 1980).

With soft automation, the set of possible operations is not determined by the configuration of the machine, but by the limits of the available programs. NC and CNC are examples of such automation. The distinction between hard and soft automation can be seen in the differing roles of Swiss type automatic screw machines and CNC lathes. Swiss type automatics are special purpose lathes with automatic bar stock feeding, designed to produce minute rotational parts in very large production runs, normally in excess of 100 000. In effect, they are dedicated lathes, designed to operate in a mass production environment. CNC lathes, on the other hand, can be reprogrammed to machine a wide variety of parts automatically and are economical for small batch production.

The automation of manufacturing was not solely concerned with the production process itself. In fact, the earliest manufacturing applications of computers were primarily in the administrative and financial areas of the factory. Systems such as payroll and invoicing were among the first computer applications. The fact that these transaction oriented procedures were already well understood greatly facilitated their early automation through computerization. An important example was the first material requirements planning (MRP) system, which linked a bill of material processor to files describing product structure and inventory status, in order to automate the function of planning material requirements.

These developments in manufacturing automation were concerned with either one individual machine or one specific function within the organization. Hence the term, **point automation**. There was no explicit strategy to integrate these point solutions or make them accessible to other point solutions. With the introduction of affordable and more powerful computers in the 1970s, the possibility of tackling larger problems presented itself. This led to the expansion of these point solutions to more integrated islands of automation.

2.4 Islands of automation

Islands of automation represent automated integrated sub-systems within the factory. The initial point solutions in the previous phase were expanded to tackle ancillary or adjacent functions. Instances of such functional expansion include the emergence of production management systems, such as Manufacturing Resource Planning (MRP II), integrated material handling and storage systems, flexible manufacturing systems and various computer aided engineering systems.

The initial point solutions, which were in place to plan Material Requirements (MRP), were expanded to MRP II systems. These systems had a set of modules to manage the full range of production and inventory management functions in a manufacturing plant, and all were built around a common set of files or database. MRP II will be dealt with in detail in Part II.

Within the production process, developments such as Flexible Manufacturing Systems (FMS) and Direct Numerical Control (DNC) represent the integration of some point automation solutions into islands of automation. For example, DNC describes a system where a number of machines in a production system are controlled by a single computer in real time through direct communication. The individual machines in the system are NC or CNC machines and the part programs required to machine a particular component are down-loaded on request from the DNC computer to the controller of the individual machine.

Islands of automation represent the current state of manufacturing integration. The proliferation of such locally expanding islands of functional automation has given rise to the CIM problem. These islands are now beginning to overlap and compete with each other. Before proceeding to a discussion of CIM in Chapter 3, the rest of this chapter will be devoted to describing several of the important islands of automation seen in industry today. This description covers computer aided process planning, automated storage and retrieval systems, robotics, macro-planning systems for manufacturing, computer aided design (CAD), and flexible manufacturing systems. Production management is left to later sections of the book. This discussion begins by presenting a framework for examining the various roles that computers play in manufacturing.

2.5 The role of computers in manufacturing

Various schemes can be used to categorize the role of computers in manufacturing. One such scheme is to consider the nature of the computer interface to the production process. This interface may be **indirect**, with the computer's role being that of an information and decision support system, without any capability to sense the process directly. In this case, the computer system manipulates information that humans have extracted from the manufacturing process and fed into the computer. Alternatively, the interface between the production process and the computer may be **direct**, with the computer directly monitoring and actively controlling sections of the manufacturing process.

Examples of direct applications of the computer include CNC, DNC and robotics. On the other hand, Computer Aided Process Planning (CAPP), computer assisted numerical control programming and computerized production management are typical examples of indirect applications.

	Indirect application	Direct application
Plant level	Macro planning models Accounting systems Production management systems Computer aided design	Computer aided warehousing Direct numerical control Flexible manufacturing systems Automatic storage and retrieval system
Operation level	Computer aided process planning Computer aided work measurement Computer aided NC programming	Computer aided testing Computer numerical control Computer based automatic assembly machines Robots

Figure 2.1 Role of computers in manufacturing.

The distinction between direct and indirect applications of the computer in manufacturing is useful in that it serves to help illustrate a common misunderstanding about the potential of the computer in manufacturing. Gunn (1982) points out that:

> 'The opportunities for mechanization in the factory have been greatly misunderstood. The emphasis has been almost exclusively on the production process itself, and complete mechanization has come to be symbolized by the industrial robot, a machine designed to replace the production operator one by one. Actually the direct work of making or assembling is not where mechanization is likely to have the greatest effect. Direct labour accounts for only 10 to 25% of the cost of manufacturing, . . . The major challenge now, and the major opportunity for improved productivity is in organizing, scheduling and managing the total manufacturing enterprise, from product design to fabrication, distribution and field service.'

A second scheme of classifying the role of computers in manufacturing is to distinguish between applications at the **plant level** and those at the **operation level**. Applications at plant level include computerized production management, computerized financial and accounting systems and Automatic Storage and Retrieval Systems (AS/RS). Applications at the operations level include CNC machines, computer supported work measurement systems, computer based semi-automatic and automatic assembly equipment, for example VCD (Variable Centre Distance), DIP (Dual In-line Package) and robotic equipment for inserting components into printed circuit boards in the electronics assembly industry.

Using these two axes, namely the nature of the computer interface to the process and the level of application, a matrix of computer applications in manufacturing can be drawn up. This matrix is presented in Figure 2.1.

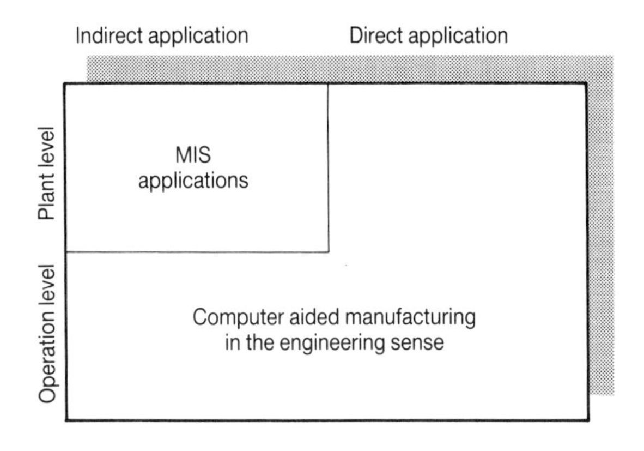

Figure 2.2 Traditional responsibilities for CAM.

An interesting aspect of this framework of computer applications in manufacturing is that it also roughly defines the traditional responsibility within manufacturing plants for various computer based applications. This is illustrated in Figure 2.2 (CAD is clearly an exception). Traditionally MIS (Management Information Systems) or DP (Data Processing) departments have had responsibility for indirect applications at plant level, while engineering groups, both product and manufacturing, have assumed responsibility for the other areas. This has led to the frequent use by engineers of the term 'Computer Aided Manufacturing' (CAM) to mean only those manufacturing areas in which engineers are involved. Perhaps this explains a frequently met view that integration is achieved when CAD/CAM integration has been realized. Seen from the perspective of the above framework, this is a narrow engineering viewpoint.

Rosenthal and Ward (1986) support this point by commenting that in many companies there are two groups:

> '(1) computer aided manufacturing (CAM) and (2) integrated manufacturing information systems for planning and control, commonly called manufacturing planning and control systems. An understanding of the nature of these sub-fields, as well as the requirements in combining them to form CIM, raise a number of . . . key implications for management. . . . CAM and manufacturing planning and control systems have different organizational roots and technical orientations. . . . Further, CAM and manufacturing planning and control systems are beginning to overlap in many manufacturing organizations . . . the traditional differences between MIS and CAM initiatives are beginning to be felt. . . . Top management must begin to encourage an integrated response despite these traditional differences.'

Leaving this discussion aside, and returning to the matrix of Figure 2.1, which included many islands of automation, we will now briefly examine some of the more important of these. Given the significant investment that these applications represent, it follows that the CIM design process must involve the recognition that these applications will not be discarded. Because of its special significance for CIM, the concept of flexible manufacturing will be discussed in greater detail in Section 2.6.

Macro planning models

At the boardroom level, corporate planning models have been used for some time. From a manufacturing point of view, these models have been developed to allow top management to understand the impact of changes in key manufacturing variables, including, for example, the level and mix of output, labour skills, new product introduction rates and levels of productivity. These models tend to represent the manufacturing system at a highly aggregated level. They can be implemented in distributed spread sheet tools. Special purpose modelling languages and packages are also available, for example, systems such as the systems dynamics approaches of Jay Forrester (1961), Roberts (1978) and Lyneis (1980). Another option is represented by fourth generation programming languages, but these still tend to be programmers' tools rather than end user tools. Current research in this area includes the application of knowledge based approaches to this problem (see Harhen *et al.* (1987), for example).

Computer aided process planning

CAPP or computer aided process planning is a computer application which supports the development and creation of the technological plan required to produce a given part. It is an important application from a CIM perspective since it is a key interface between Computer Aided Design (CAD) and Computer Aided Manufacture (CAM). The resulting process plan consists of a statement of the sequence of operations necessary to manufacture the part, the identification of the machines on which these operations should be carried out, as well as the operation times. Special tooling and set-up procedures are also identified. To date, the majority of the work done on CAPP has been oriented towards metal cutting applications.

There are two approaches to developing CAPP systems:

(1) The variant approach.

(2) The generative approach.

The variant approach involves the preparation of a process plan through the manipulation of a standard plan or the plan of a similar part. The process plan for the master composite part is stored in the computer and used in the planning of subsequent parts. The master part is a composite of all features likely to be seen on the parts to be planned. The variant approach uses parts classification techniques to identify features on parts and match them to equivalent features on the master part.

The generative approach involves the creation of the process plan from information available in the manufacturing database. This approach requires a detailed description of the part to be planned, the various manufacturing operations available and the capabilities of these operations in terms of process accuracy, tolerances, etc. Thus, for example, in the context of a machining application, the system looks at each surface to be machined and compares the surface tolerance with that achievable using an available process. If the tolerance can be achieved by the process, then that process may be selected to produce the surface. If not, it is eliminated from further consideration.

In recent times, researchers have begun to prototype CAPP systems using Artificial Intelligence (AI) techniques, for example, the work of Descotte and Latombe (1981, 1985) and Bowden and Browne (1987). CAPP has many features that make it an appropriate application area for knowledge based systems. The process planner uses knowledge of the various manufacturing processes, machines and tools required to carry out particular operations, as well as knowledge in the form of experience gained from planning previous parts. AI tools provide the capability to represent such knowledge efficiently and control its application to the process planning problem.

Automatic storage and retrieval systems (AS/RS)

An AS/RS (sometimes referred to as an automated warehouse) is a system which stores and retrieves materials using automatic stacker cranes under computer control. The system identifies each arriving pallet, typically using bar code technology, selects an appropriate open location on the storage racks and directs the stacker crane to route the pallet to that location. When a request is made to retrieve a stored pallet, the computer identifies the storage location and directs the stacker crane to the specified location for retrieval. Rygh (1980) claimed the following benefits for an AS/RS over conventional warehousing methods:

- improved floor space utilization,
- reduced direct labour cost,
- virtually 100% accuracy in inventory measurement,
- reduced pilferage,
- lower energy consumption,
- reduced product damage,
- improved customer service.

Computer aided design (CAD)

CAD involves the use of computer hardware and software to assist the designer in the storage, manipulation, analysis and reproduction of his/her design ideas. Modern engineering CAD systems allow designers to produce 3D (three dimensional) geometric models. These geometric models may be constructed using wireframe models, surface models or solid models.

Wireframe models represent the edges of surfaces by lines. Surface models allow the faces of a part to be built up so that a part can be represented by all of its faces in the form of a shell. Solid models allow the true 3D geometry of a part to be represented, in the sense that the solid model is geometrically and mathematically complete. Complete means that the designer can calculate the mass of the design, determine its volume, check clearances and tolerances, determine the centre of gravity of the part, calculate its moments of inertia, and even take sections through it. As might be expected with this rich functionality, solid modellers are very expensive and require significant computer resources.

CAD in the electronics industry seems to have reached a more advanced stage of integration than in the mechanical engineering world. Integration from logic design and simulation, through circuit board and chip layout and onto post-processing for assembly programs and test programs, has been achieved to a significant extent in state of the art VLSI (Very Large Scale Integration) CAD/CAM installations.

Robotics

A robot is a reprogrammable multifunctional device, designed to move material, parts, tools or specialized devices through variable programmed motions, for the performance of a variety of tasks. Robot applications can be found in virtually all branches of industry, particularly in the automotive, electrical, electronic and mechanical engineering industries. The major application areas across these industries are spot welding, arc welding, surface coating (including spray painting) and servicing machines which includes the loading and unloading of machine tools, die casting machines, forging presses and plastic injection moulding machines. To date, there have been few applications of robots in small parts assembly.

A survey (Sanderson *et al.* 1982) designed to determine the basis for installing robots in industry indicated the following reasons in decreasing order of importance:

- To reduce labour cost.
- To replace people working in dangerous or hazardous conditions.
- To provide a more flexible production system.
- To achieve more consistent control of quality.

- To increase output.
- To compensate for a local shortage of skilled labour.

The present generation of robots is more or less restricted to working in highly structured environments. Their flexibility is extremely limited in that they are unable to deal with their environment in an intelligent way. For this reason, the vast majority are involved in relatively primitive, simple and repetitive manufacturing tasks. The emerging generation of robots is *intelligent* in that they have built-in sensor systems that allow them to react to limited changes in their environment. The use of intelligent sensors, such as vision or tactile sensors, means that robots which, to date, have been limited to performing structured, highly repetitive operations, will be better able to react to changes in their working environments. However, present day industrial robots technology is still at a very early stage of development and the robots portrayed in fiction with advanced levels of mobility and intelligence are still technologically and economically infeasible.

2.6 Flexible manufacturing systems

Because of the special significance of Flexible Manufacturing Systems (FMS) within CIM, FMS will be dealt with at some length in this section.

A flexible manufacturing system is an integrated computer controlled system of automated material handling devices and CNC machine tools, that can simultaneously process medium sized volumes of a variety of parts. An FMS, therefore, incorporates many different automation concepts into one system. These include:

- Numerical Control (NC) and Computer Numerical Control (CNC).
- Robotic process equipment.
- DNC control of the material handling system and the individual CNC machines.
- Automatic material handling.
- Automatic tool changing.
- Automatic machine loading and unloading.

Parts enter and leave the FMS at a central location and the material handling system automatically transports the parts to the machines identified by the process plan or routing. The routing for each part and the operations required to produce it, differ across the range of products. This routing is finally determined within the FMS by a scheduler which has access to the product data and the requirements schedule. It can thus determine on which machines a part should be processed next. Knowing this, the scheduling algorithm or heuristic generates the required schedule.

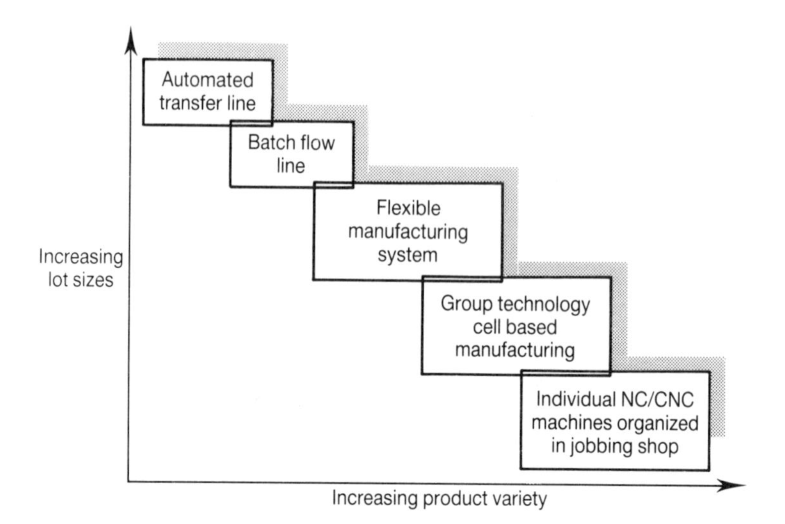

Figure 2.3 The applicability of FMS.

The control of the FMS is exercised by a computer, or perhaps a hierarchy of computers. The control system is responsible for scheduling work within the FMS and coordinating the material handling system and the machines to meet this schedule. The programs required to produce the parts are down-loaded to the individual CNC machines. Orders are transmitted to the materials handling system, detailing which part should be moved to each location. The computer also has the responsibility for producing reports on the operation of the FMS. Data is gathered from each of the machines and this is collated into system reports. Hierarchical scheduling systems for flexible manufacturing systems are described by Akella *et al.* (1984).

The development of FMS is important because, in a sense, an FMS can be considered a mini-CIM system. FMSs are designed to fill the gap between high volume hard automation transfer lines and CNC machines. On the one hand, transfer lines are very efficient at producing parts in large volumes at high output rates, with the important limitation that the parts be identical. These highly mechanized lines are very inflexible and will not tolerate variations in part design. The CNC machine, on the other hand, offers great inherent flexibility through its ability to be reprogrammed to machine parts of varying contours and sizes. The FMS lies somewhere between these two extremes as indicated in Figure 2.3.

Furthermore, the term FMS covers a range of systems including single machine Flexible Manufacturing Cells (FMC) through to Flexible Transfer Lines and the FMS itself, which consists of a group of FMCs and/or CNC machines connected through an automated materials handling system.

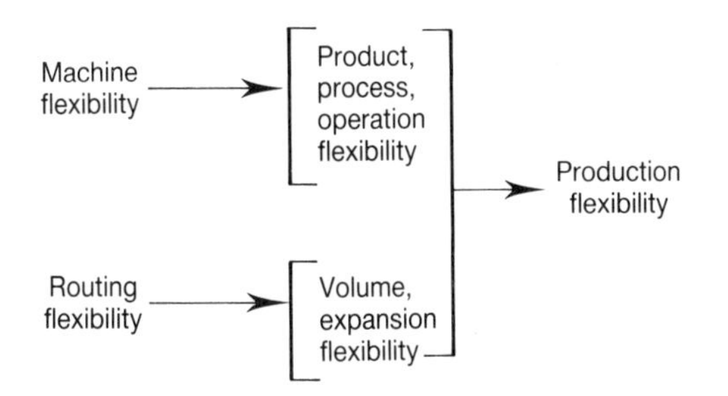

Figure 2.4 Relationship between flexibilities.

(See Stecke and Browne (1985) for further details). The definition of flexibility in the context of FMS is important. Browne *et al.* (1984) identified and defined eight types of flexibility in the context of FMS.

(1) Machine flexibility refers to the ease of making the changes required to produce a given set of part types.
(2) Process flexibility refers to the ability to produce a mix of jobs.
(3) Product flexibility refers to the ability to change over to produce new products very economically and quickly.
(4) Routing flexibility refers to the ability to handle breakdowns and to continue producing a given set of part types.
(5) Volume flexibility refers to the ability to operate the FMS profitably at different production volumes.
(6) Expansion flexibility refers to the capability of expanding a given FMS, as needed, easily and in a modular fashion.
(7) Operation flexibility is the ability to interchange the ordering of several operations for each part type.
(8) Production flexibility refers to the universe of part types which the FMS can produce.

Clearly some of these types of flexibility are related (see Browne *et al.*, 1984). Figure 2.4 indicates the dependency relationship among the various flexibility types. The arrows in this figure signify the relationship *necessary for*.

Butcher (1986) describes an FMS for the production of aero-engine disc components, installed at Rolls-Royce in the UK. The objectives set for the FMS were to 'cut work in progress by two thirds, compress production lead times from 20 to 6 weeks and increase manpower productivity by over 40%'. The installation of the CNC machines for the FMS reduced the number of operations from 21 to 5. The machine tool population within the plant was halved from 62 to 31 machines and the scrap rate was reduced by 40%. Westkamper (1986) describes the results of FMS installations at one of MBB's aircraft production plants in the Federal Republic of Germany. Lead times were reduced by 25%, floor space requirements by 42%, the number of machines by over 50% and personnel by over 50%. It is clear, even from these two examples, that the characteristics of FMS are somewhat similar to the likely characteristics of the *factory of the future*, namely, a small number of multifunction production machines, low WIP (work in progress), short lead times and very low labour content. These characteristics will be discussed in more detail in Chapter 3.

2.7 Conclusion

In this chapter, we reviewed the various stages of the application of automation to manufacturing. On the way, it was pointed out that a holistic approach is necessary in the design of CIM systems. Brief descriptions were presented of some of the important islands of automation to be seen in manufacturing firms today, with particular emphasis on flexible manufacturing systems. This has all served to act as a foundation for the discussion of computer integrated manufacturing, which follows in Chapter 3.

CHAPTER THREE

Computer integrated manufacturing

3.1 Introduction

Computer Integrated Manufacturing (CIM) represents the integrated application of computer technology to manufacturing in order to achieve the business objectives of the firm. CIM is being seized on by manufacturing firms as the weapon with which to achieve competitiveness. In this chapter, some of the emerging characteristics of the CIM environment are identified. The goal is not to specify a CIM architecture but to describe, instead, various viewpoints of CIM. In this manner a position will be approached from which the role of a production management system within CIM can be discussed. Readers interested in a more technical discussion of CIM from an architectural point of view are referred to such sources as the ESPRIT (European Strategic Programme for Research and Development in Information Technology) publications, for example, Yeomans (1985) and the work done within the CIM OSA (CIM Open Systems Architecture) project of ESPRIT.

This chapter initially focuses on the nature of the factory of the future. Some time is then devoted to discussing CIM from various functional perspectives. Finally, an enterprise-wide view of CIM is described under the heading 'Computer Integrated Business' (CIB).

3.2 The factory of the future

What will the factory of the future look like? Some likely characteristics are as follows:

- **Round the clock operation** The high capital cost of equipment will require that the equipment work more or less continuously to pay for itself.

- **Very small lot sizes** Products and components will be manufactured and assembled in small lots. In fact, lot sizes will approach one. Small lot sizes offer great flexibility in times of rapid product design change and changing customer requirements.
- **Greatly reduced lead times** The cycle time from placement of an order to delivery to the customer will be measured in days and hours, not months and weeks. This will involve achieving significantly reduced in-process lead times, as well as carefully managed inventory policies. The reduction of cycle time is probably the key characteristic of the CIM factory from a business point of view.
- **Little or no human labour at the point of production** The number of people directly involved in manufacturing is constantly decreasing. In the factory of the future, the emphasis will be on the capabilities of human labour as distinct from its cost. At a societal level, predictions have been made by professional organizations, such as the Society of Manufacturing Engineers in the USA, that the numbers engaged in direct labour in manufacturing in the USA by the beginning of the next century will represent a very small percentage of the work force. This has important social consequences which are beyond the scope of this book.

  The reduced significance of labour as a factor of production also has important ramifications for the methods used to allocate cost in manufacturing systems. Labour has traditionally been the focus of cost control. When direct labour costs are of the order of 1% of the total cost of the product, innovative manufacturing cost management systems will need to be put in place. Spur (1984) talks about the factory of the future involving huge investments and the products produced in it having very high fixed costs and low variable costs.
- **The factory of the future will be small** It will almost certainly be composed of a small number of extremely versatile machines which are integrated to look almost like a single machine. We already have a glimpse of this future when we look at computer numerically controlled machining centres, which are highly versatile metal cutting machines with automated tool changing and automated pallet (of work) loading and unloading. Talaysum *et al.* (1986) have termed such versatile machines 'multi-mission production facilities'. Weston *et al.* (1986) talk about distributed manufacturing systems being composed of flexible *intelligent* machinery linked by distributed software systems. The development of standard factory communication protocols at the application level, e.g. the MAP (Manufacturing Automation Protocol) development, as well as the strengthening network capabilities provided by some of the computer systems manufacturing companies, will facilitate communications between intelligent machines.

Spur (1984) presents the concept of a highly decentralized manufacturing system with individual manufacturing units, which are geographically dispersed, 'equipped with all of the devices and instruments necessary to carry out a partial or complete processing of groups of parts', while 'those services and departments which cannot be decentralized remain in the main plant, e.g. purchase, storage, production planning, main production control etc.' These decentralized units might well be 'independent small businesses cooperating with the (central) factory on a long term basis'.

3.3 The nature of CIM

CIM is the technology which will realize the *factory of the future*. If one considers the characteristics of the factory of the future as identified above, one may be reminded of today's highly automated and integrated continuous process industries. Indeed, it can be argued that CIM seeks to achieve in discrete parts manufacturing, the type of integration already achieved in many continuous process industries, such as steel making and oil refining.

There are many reasons for the emergence of CIM. Ultimately it is driven by the desire to apply ever increasing levels of automation to the manufacturing system in order to attain improved productivity. Japanese competition has spurred on the Western industrial nations to devote great efforts to achieve this. The ever decreasing cost of computer technology has accelerated this process.

Nowadays, existing islands of automation are beginning to overlap and compete with each other for the right to perform certain tasks within the factory. For example, a manufacturing plant considering the implementation of a Manufacturing Resource Planning system (MRP II) and an Automated Storage and Retrieval System (AS/RS), will typically deal with two different vendors. Moreover, plant inventory status can be monitored by either or both of the two systems. The user of this technology is faced with the need for adequately defined communications between the systems, or else has to live with some degree of duplicate data entry and all of its consequences in terms of effort and possibility of error.

Interfacing islands of automation which have been designed in isolation will always be problematic. The research programmes into CIM are attempting to alleviate these problems by redesigning the modules of the manufacturing system with a view towards full integration. One such project is ESPRIT project 477 (Actis-Dato *et al.*, 1986).

The following are just some of the benefits which will accrue from CIM.

- reduced inventory investment,
- reduced manufacturing lead time,
- reduced design cycle time,

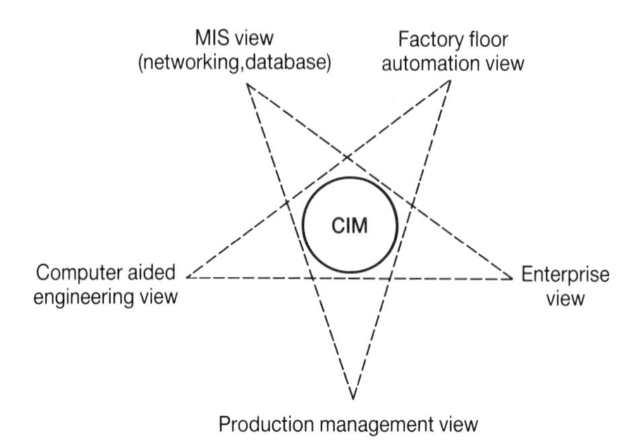

Figure 3.1 Different perspectives on CIM.

- improved utilization of plant and labour,
- improved control of the total manufacturing system.

Computer integrated manufacturing involves the integration of the various islands of automation into a coherent whole. Although a number of definitions have been put forward, no strong agreement has been reached on the scope of CIM. CIM is an industry driven technology, with each branch of industry conditioned by its own particular set of experiences, requirements and circumstances. Furthermore, it is clear that an industrial company seeking to realize CIM, will have to take account of its own existing investment in islands of automation, before planning an evolution towards CIM. Because of all of these circumstances, descriptions of CIM tend to be highly dependent on the background and perspective of the individuals offering them. The following are some examples of statements describing CIM.

> 'A closed loop feedback system whose prime inputs are product requirements and product concepts and whose prime outputs are finished products. It comprises a combination of software and hardware, product design, production planning, production control, production equipment and production processes.'
>
> Merchant (1977)

> 'The logical organization of individual engineering, production and marketing/support functions into a computer integrated system. Functional areas, such as design, inventory control, physical distribution, cost accounting, planning and purchasing, are integrated with direct materials management and shop floor data acquisition and control. Thus the loop is closed between the shop floor and its controlling activities. Shop floor machines serve as data acquisition devices for the control system and often its direct command . . .'
>
> Anon (1981)

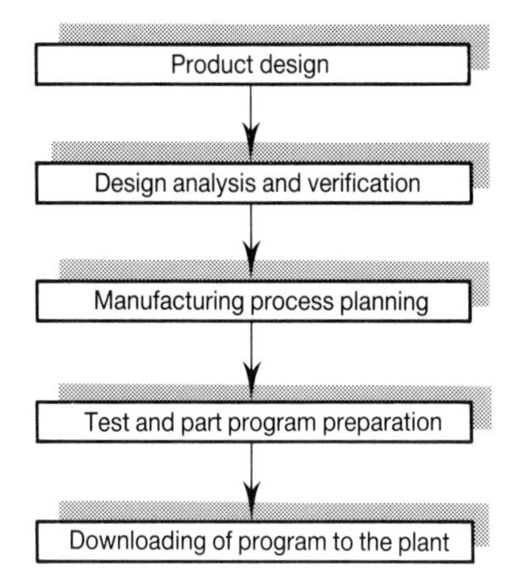

Figure 3.2 CAE view of CIM.

> 'Computer integrated system involving the overall and systematic computerization of the manufacturing process. Such systems will integrate computer aided design, computer aided manufacture and computer aided engineering, testing, repair and assembly by means of a common database.'
>
> ESPRIT (Commission of the European Communities 1982)

> 'The integrated application of computer based automation and decision support systems to manage the total operation of the manufacturing system, from product design through the manufacturing process itself, and finally on to distribution; and including production and inventory management, as well as financial resource management.'
>
> Harhen and Browne (1984)

In Figure 3.1 some of the differing perspectives offered on CIM are presented.

Consider a pure engineering perspective. Engineers tend to see CIM primarily in terms of CAD/CAM integration, as illustrated in Figure 3.2. CIM from this perspective is dominated by the Computer Aided Engineering (CAE) task.

CAE has a view of CIM that is a partial reality in some industries today, e.g. VLSI fabrication and, to a lesser extent, electronics assembly. There are greater difficulties in realizing CIM in the mechanical engineering industry, where the major difficulty is the interface between CAD, CAPP and NC programming. Much research is underway to develop systems that overcome these problems, integrating, for example, CAD and robot programming systems.

In the domain of CAE, there is increasing awareness of the need for *design for production and assembly*. There is a consequent necessity for the

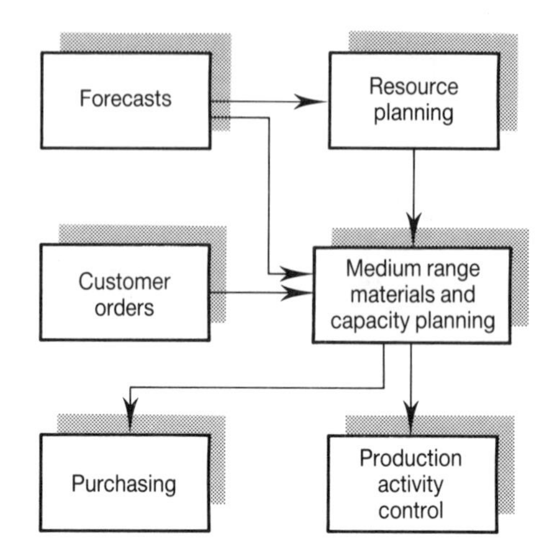

Figure 3.3 PMS view of CIM.

manufacturing function to influence the design process and to ensure that the designers are aware of the effects of various design features on the ease of manufacture of a part. A good description of the issues is contained in Boothroyd and Dewhurst (1983).

In the medium term, the benefit of an integrated CAD/CAM facility is the reduction in the lead time from initial design concept to the manufactured product. This greatly reduced design cycle provides industry with the facility to respond quickly and economically to changes in the market place.

If CIM is viewed from the point of view of the Production Management System (PMS), a very different picture is obtained, as illustrated in Figure 3.3. The system is essentially seen as a hierarchy of scheduling systems. The resource planning system is concerned with setting long range aggregate production programmes and resource needs. Medium range materials and capacity planning is typically an information systems function that converts the master schedule into a more detailed plan for the medium range, using appropriate batching techniques for planning both production and material acquisition. Material Requirements Planning (MRP) is a good example of such a system. The management of the satisfaction of actual customer orders is achieved at this level as well. The lowest level in the production management system involves the execution systems, such as the production activity control system and the purchasing system.

The Production Activity Control (PAC) system plays a particularly important role within the CIM system. The PAC system manages the flow of material and the associated flow of data through the plant (see Figure 3.4). Essentially, it is through the PAC system that the PMS is linked to the factory floor.

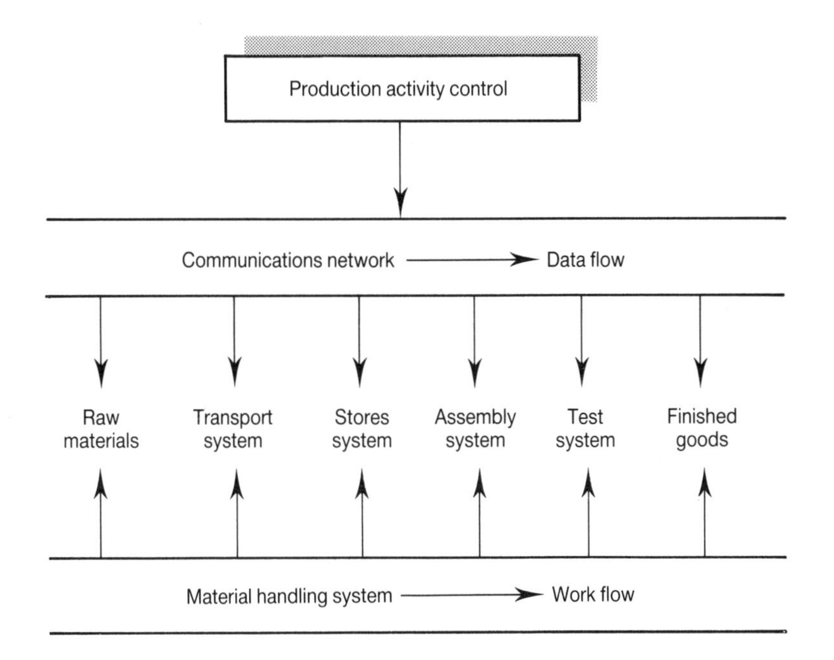

Figure 3.4 PAC and the factory floor.

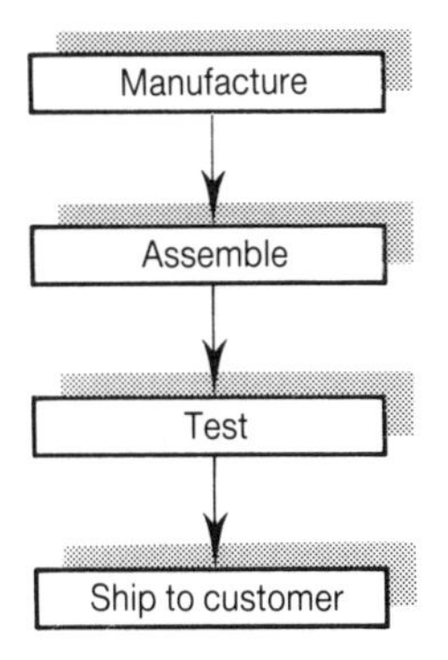

Figure 3.5 Material flow view.

If CIM is viewed from the point of view of the material flow, then the CIM issue becomes one of designing appropriate material flows across the various stages of manufacturing. This involves the specification of the manufacturing process, as well as the provision of effective means to execute the material movement, as illustrated in Figure 3.5.

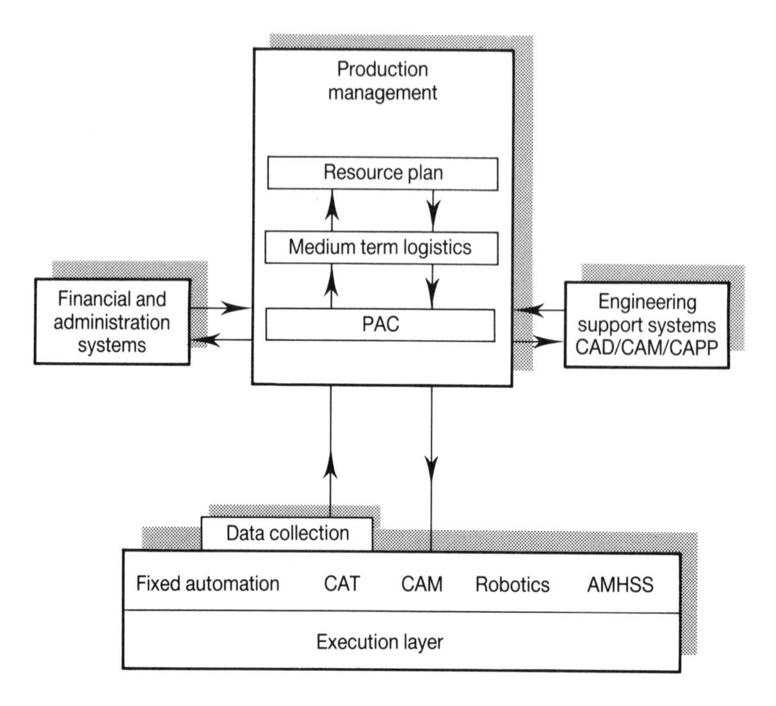

Figure 3.6 CIM and its constituent functions.

Figure 3.6 presents a unified view of CIM. The connection with sales and marketing occurs through the master planning and order entry functions within production management.

The lines of communication highlight the integration of these functions and the closed loop feedback through data collection to production management and other systems. In summary, computer integrated manufacturing is the functional integration of the following functions, served by computer communications and data storage facilities:

- **Administration and financial systems** The administration and financial systems cover order invoicing, long range planning and budgeting within the organization. These systems are concerned with the costing of production and materials, and control of administrative and financial aspects of the firm.
- **Engineering support systems** Engineering and support systems are concerned with the design of products and the development of processes. Included are Computer Aided Design (CAD), Computer Aided Engineering (CAE) and Computer Aided Process Planning (CAPP).

- **Production management** The production management function coordinates manufacturing related activities in order to achieve an appropriate balance between the goals of customer service, process efficiency and minimum inventory investment.
- **Execution layer systems** The execution layer systems are those computer based applications which directly affect the execution of production. Examples are Computer Aided Test (CAT) and Automated Material Handling Systems (AMHS).

It is no coincidence that the PAC system is represented as the heart of the CIM system. This, in the authors' opinion, is the key function within CIM at the operational level. It represents the gateway between the planning functions of PMS and CIM, and the manufacturing process. Plans flow downward through this gateway and data detailing progress flows upwards. Because of its central importance in the realization of CIM at the operational level, Chapter 4 devotes considerable effort to describing the role of the PAC system within the CIM environment. The remainder of this chapter is devoted to describing a broader functional view of CIM.

3.4 Computer integrated business

CIM is regarded as a system which represents the highest level of integration between the various manufacturing functions in the firm. However, implicit in this discussion of CIM has been the idea that CIM is a *four walls* activity. This focused view of CIM does not directly consider integration with outside systems, such as customers, sales or suppliers. For example, sales personnel must know what products are being produced by manufacturing and must also have an accurate estimate of the length of time it takes to procure products. In larger enterprises, a more complex situation exists in which interactions between the several different levels of manufacturing plants must also be managed.

A system can therefore be visualized wherein functions, such as sales, marketing and vendors, are all integrated with the factory CIM system by means of an appropriate set of applications. This view of CIM as the integration of all these external systems with the factory CIM system is called **Computer Integrated Business** (CIB). The achievement of CIB depends, to a large extent, on the development of protocols and applications to support Electronic Data Interchange (EDI). Figure 3.7 represents a single plant CIB system.

CIB is concerned with reducing the cost and time taken to transfer information between the factory and the external systems with which it must interface. For example, at the front end of the business it can take a significant amount of time to process the order from the customer and later to

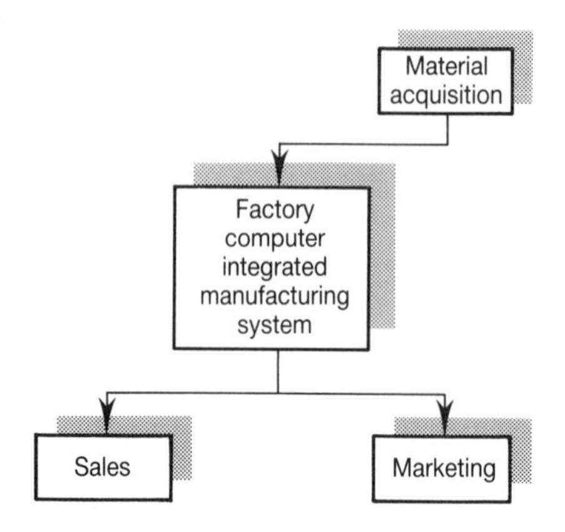

Figure 3.7 Computer integrated business.

manage the shipment of the order. At the back end of the business, the transaction processing time in acquiring the raw material from the vendors is also significant.

In some industries, the lead time and cost of manufacturing is less significant than the transaction cycle times and costs associated with processing customer orders, acquiring raw material and distributing finished product. This is so for a number of reasons, primarily because of the problems associated with product diversity and customized products. For example, in the purchase of complex products, such as computer systems, the technical verification of an order as a valid configuration is quite a complex task. Knowledge based approaches have been shown to solve this problem successfully. The pioneering work in this area was the XCON project of Digital Equipment Corporation and is described by McDermott (1981) and Bachant and McDermott (1984). An extreme example of the relative complexity of transaction processing, as opposed to manufacturing, is met in the case of volume manufacturing of complex software and documentation.

CIB can thus be seen as taking a wider view than CIM in its management of the various cycle times associated with manufacturing. This is depicted in Figure 3.8. As the diagram shows, CIM is concerned with activities in the manufacturing phase of an order whereas computer integrated business is concerned with all aspects of the relationship with the customer, from receipt of initial order to dispatch of product and, in certain cases, with the post-order relationship with the customer to cover maintenance and product update.

In the discussion thus far, CIB is described for the single plant company. In many large manufacturing companies, systems of multiple plants must be coordinated to produce finished products. CIB systems,

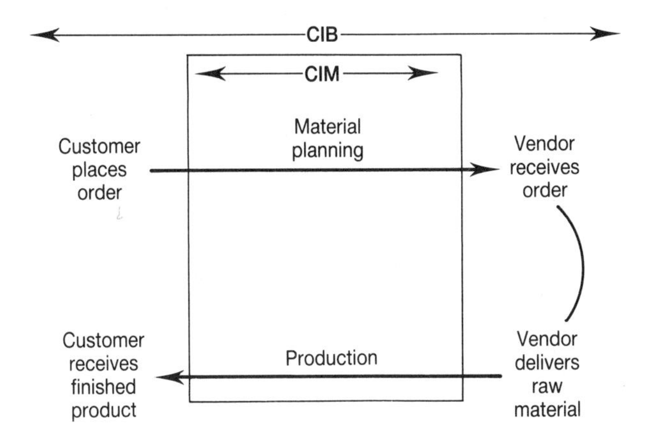

Figure 3.8 The scope of CIB.

therefore, need to be able to manage the global nature of modern manufacturing enterprises. This, of course, makes the CIB problem much more complicated.

In such a situation, a plant may either be a feeder plant or a header plant, which links directly with the customer. There are thus supplier/customer relationships between many of the plants in the manufacturing system. However, not all of the manufacturing functions discussed previously as part of CIM exist separately in each plant. The following are some of the situations that may arise:

- The total knowledge of a product is not solely owned by a single plant but is shared with other supplier or customer plants.
- A product cannot be designed with consideration for just one plant, since the needs of alternate manufacturing plants must be taken into consideration.
- Certain functions, such as master planning and order entry, are not local independent activities but must be connected with related functions within all other plants in the corporation.
- Alternatively, such functions may be centralized in one location and so may not necessarily reside in any of the plants. In such cases, there still remains a problem of how this centralized function should communicate with the plants.

From these observations it can be seen that four walls CIM in these plants would not be as functionally rich as in a single plant company. However, the integration with the business functions in the rest of the firm will be a much more complex activity. Figure 3.9 represents the multiplant case of CIB.

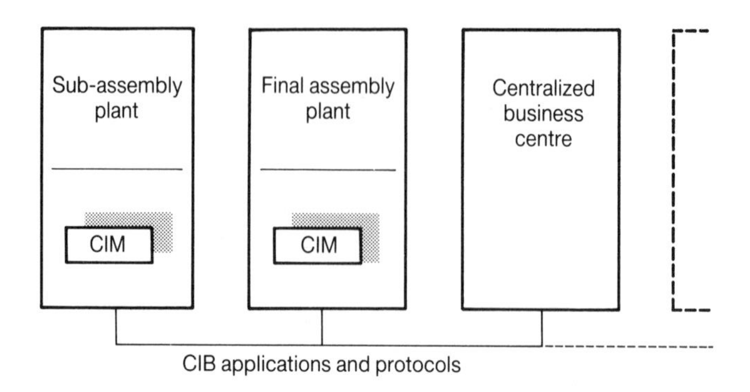

Figure 3.9 Corporate-wide CIB.

To summarize, CIB represents the integration of the factory CIM sub-systems with those business functions outside the plant, with which the factory must interface.

3.5 Conclusion

In this chapter, computer integrated manufacturing has been described. This description began with an account of the characteristics of the factory of the future. The reasons for the emergence of CIM were covered. Various descriptions and definitions of CIM were examined before presenting a composite view. CIM as computer integrated business was also described. This was done both from a single plant and multiplant perspective. In Chapter 4, the production management system will be discussed, with particular emphasis on the PAC system and its role within CIM systems.

CHAPTER FOUR

The role of production management in CIM

4.1 Introduction

The Production Management System (PMS) lies at the heart of the CIM system. It regulates the pulse of the manufacturing system at the operational level through its decisions of what and when to buy and make. This chapter is devoted to describing the role of the PMS within CIM. Initially, we shall look at the hierarchical nature of manufacturing planning. Secondly, we shall explore the role of the manufacturing process organization and its impact upon the nature of the production management system. Finally, as has been seen, the PAC system is the main gateway between the PMS and the rest of the CIM system. Consequently the remainder of the chapter is devoted to discussing how PAC is interfaced with other islands of automation and thus facilitates the achievement of CIM.

4.2 Production management systems

Manufacturing planning can be viewed as a hierarchical process. The hierarchy of planning and control functions in manufacturing extends from the corporate charter down to the real time control aspects of production activity control. This hierarchy is illustrated in Figure 4.1. The distinguishing features at each level are the time horizon employed and the level of detail used in representing planning information.

The corporate charter is a statement of the fundamental goals and policies of the organization and, as such, can be considered as having an infinite time horizon. This statement includes reference to the market in which the organization operates, its relationship with its employees and its policy towards growth, profit and quality.

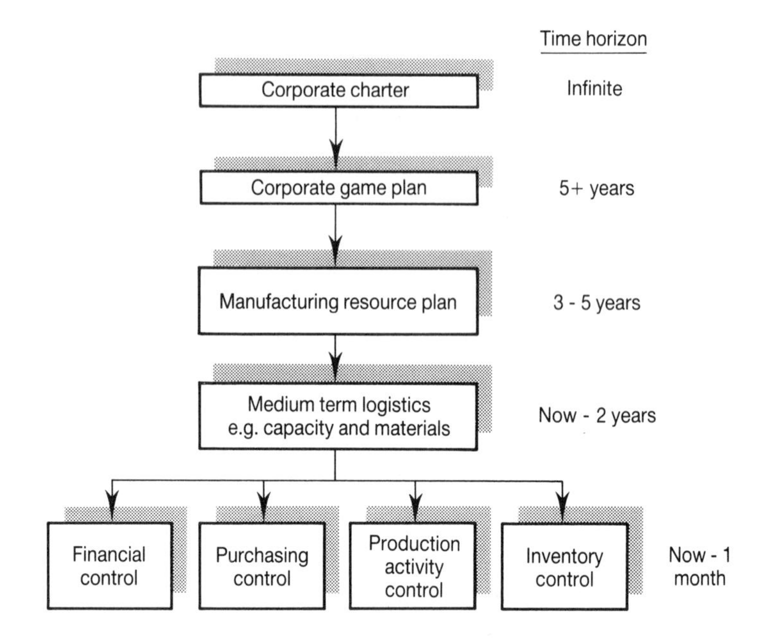

Figure 4.1 The hierarchy of manufacturing planning.

The corporate game plan is a statement of the operating goals that will guide the organization for the next five years or more. It is corporate by nature. Included are quantitative statements about product strategy, manufacturing strategy, global investment strategy, growth of sales and market share. This is developed with the aid of financial, manufacturing and marketing macro-models.

The manufacturing resource plan is a quantitative statement of requirements, in terms of people, plant, inventory and finance, for three to five years ahead. It can be plant focused or corporate focused. It is determined by the time phased explosion of a gross top level production plan. It allows for the design of hiring programmes, plant construction programmes and the planning of the provision of long term finance. The production plan that drives the manufacturing resource planning system must be in tune with the market plan over the same time horizon. Collectively, the production plan and the market plan are known as the business plan.

The next level is the medium term logistics level, which includes setting the Master Production Schedule (MPS) as well as the determination of material and capacity requirements. Much of these functions are supported by MRP II (Manufacturing Resource Planning) systems. The MPS is a detailed statement of top level planned production over the short to medium term. The master production schedule seeks to balance incoming customer orders and forecast requirements with available material and capacity. The

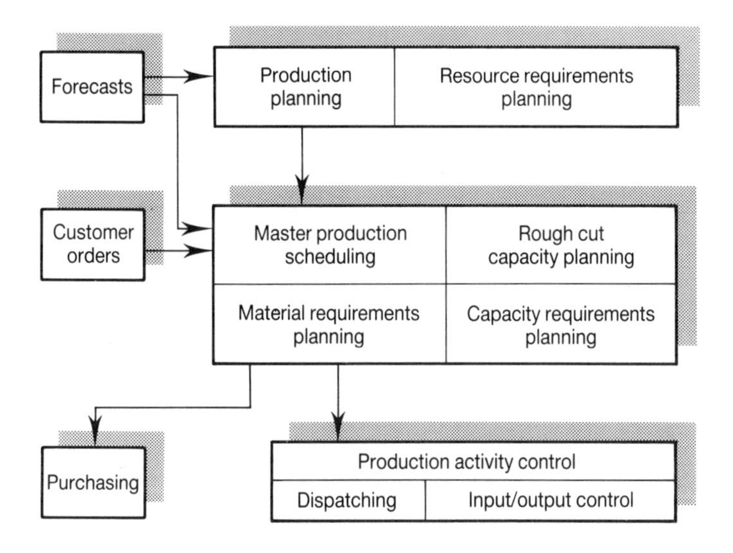

Figure 4.2 The hierarchy of PMS.

material plan is a detailed recommended schedule for both purchase and production order release. It can be calculated by applying the time phased explosion and netting off process of MRP to the MPS. The MPS and the material plan can look up to two years ahead, depending on the specific industry. The material plan feeds the various lower levels which execute the plan, making adjustments as necessary. These are financial control, inventory, purchasing and production activity control.

Not only does good planning at the higher levels facilitate control but good control can also facilitate planning, as indicated by French (1980). These various control functions are inextricably interrelated. For example, an inventory shortage caused by a purchasing problem affects production and eventually profits. The hierarchy can be further articulated for production activity control, but this is seen as beyond the scope of this book. Interested readers are referred to Harley *et al.* (1986a, 1986b).

At each level, the decision process has to maintain a balance between two opposing forces, i.e. between determining what has to be produced (priority) and whether the facilities are available to produce these (capacity). This opposing balance is clearly visible in Figure 4.2.

4.2.1 Product structure and PMS

Clearly the more components a product contains and the more levels of manufacturing involved, then the greater the complexity in controlling the production of that product. Thus production management tends to be simpler

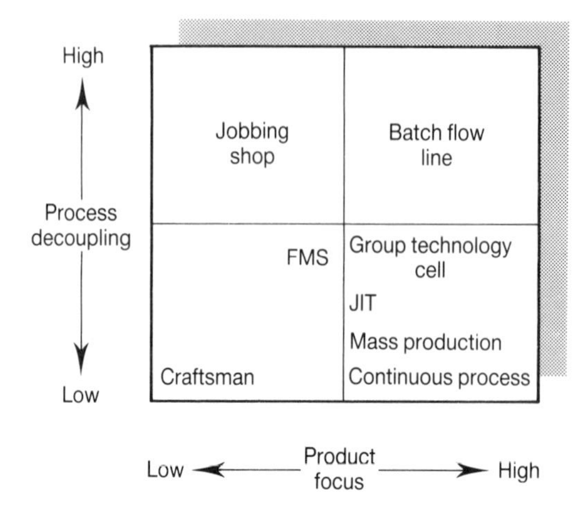

Figure 4.3 Types of procress organization.

in the case of a one piece fabricated product, such as a spanner, and more complex in the case of complex assembly products, such as automobiles or computer systems.

4.2.2 Manufacturing process organization and PMS

Manufacturing process organization has already been discussed in Chapter 1. There we identified mass production, batch production and the jobbing shop as the dominant forms. In this section, we shall develop that discussion and seek to understand how the manufacturing process organization influences the nature of production management.

A manufacturing process organization may be viewed along two dimensions:

(1) **The degree of process decoupling** To what extent is the production process for a product divided into separate operations and decoupled by inventory buffers?

(2) **The degree of product focus** To what extent are production facilities devoted to specific products?

Using this to create a matrix, we can then position various forms of manufacturing process organization on this framework. This is illustrated in Figure 4.3.

An example of low process decoupling, low product focus is a solitary craftsman who makes a variety of products. So too is the automatic engineering cell known as a Flexible Manufacturing System (FMS). In these cases there is product variety and little or no decoupling of the various stages of manufacturing.

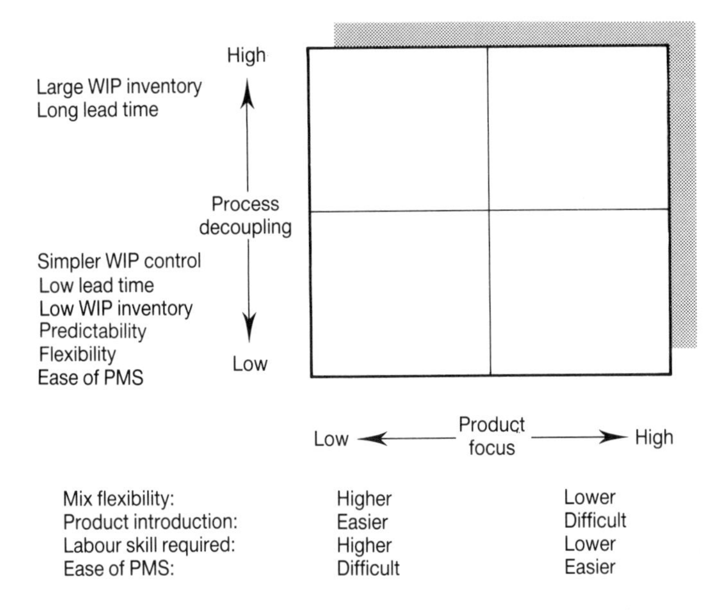

Figure 4.4 Benefits of differing types of process organization.

The traditional jobbing shop involves high process decoupling and low product focus, whereas the group technology approach involves low process decoupling and high product focus, as do continuous process plants (e.g. oil refining). Mass production with a fully automated transfer line also has the same characteristics since, although there may be a fine division of the process into many process stages, each stage is finely balanced with the next and there are no intermediate buffers to decouple the process stages. The JIT system also emulates this by removing all buffers from the process and by organizing the material flow process and the master schedule so that continuous flow manufacturing can be approached. (See Chapter 13.)

Finally, the medium to large batch flow line with many process stages falls into the high process decoupling, high product focus category. The benefits of the various types of manufacturing process organization are illustrated in Figure 4.4.

A process which is not decoupled into many stages has a short manufacturing lead time and consequently a low Work in Progress (WIP) investment. WIP control is also facilitated. It also has the most predictable delivery performance and can support flexibility at the master schedule level in response to the mix or volume changes. In addition, low process decoupling tends to encourage ownership and responsibility by the operator for good manufacturing performance as measured in product quality and delivery performance.

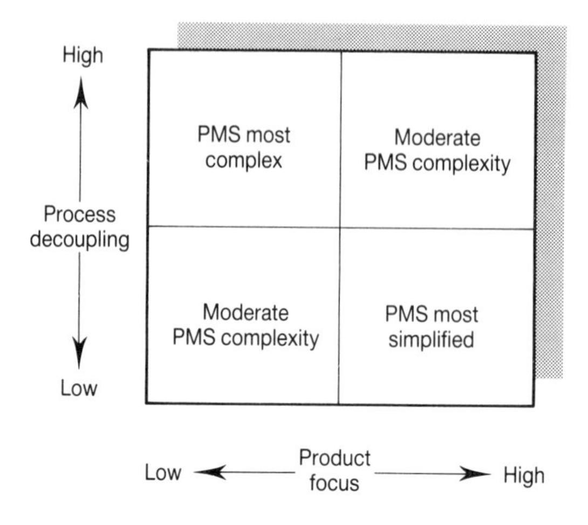

Figure 4.5 Effect of manufacturing process organization on PMS.

Where a manufacturing process is highly decoupled, high process efficiency, in terms of both labour and capital utilization, can, in theory, result. This is because the decoupling of the various stages of the process avoids problems, such as *waiting for work*, and enables set-up cost economies to be made by batch production. In addition, the physical division of the process facilitates the application of automation to the manufacturing process. It is interesting to note that the automated transfer line type of mass production does not exploit all potential efficiency gains. This is because of the so-called balancing loss in a transfer line. Process efficiency has been marginally sacrificed in order to reduce lead time and WIP inventory.

Where a process is product focused then product quality also tends to improve. Less labour skill is required. In addition, set-up economies are possible because machines are permanently set-up, or at least there is high set-up commonality. However, flexibility tends to be impacted, since there is less flexibility in the system for dealing with product variety. As a result, new product introduction is more difficult.

The complexity of PMS in each of the four quadrants of the process organization framework is illustrated in Figure 4.5. The jobbing shop is the most difficult PMS environment. In batch line production, PMS is of medium difficulty. In FMS or in the case of a craftsman, the PMS problem has lower complexity. PMS is least complex in the case of JIT, group technology and transfer line type mass production.

The above sections have described the functionality of PMS and given an indication of the complexity of PMS in different manufacturing process organizations. We shall now proceed to focus on PAC and the role that PMS plays in CIM.

4.3 Production activity control

Production Activity Control (PAC) is the layer of the PMS that lies closest to the production process. As has been clearly indicated, it plays a very important role in linking the factory floor with the other elements of PMS. It transforms planning decisions reached at the higher levels of the PMS hierarchy into control commands for the production process. Complementary to this role, it also translates data from the shop floor into information which is used to aid the higher level planning functions in the PMS. The remainder of this chapter is devoted to expanding the description of PAC. PAC can be defined as follows.

> 'Production activity control describes the principles and techniques used by management to plan in the short term, control and evaluate the production activities of the manufacturing organization'.
>
> Harhen and Browne (1984) adapted from the APICS definition.

PAC is concerned with activities ranging from firming the release of planned manufacturing orders to the analysis performed after order completion. At the order level PAC can also evaluate the effectiveness of manufacturing activities (see Harhen *et al.*, 1983). PAC thus includes all of the following PMS functions:

- manufacturing order approval and release,
- operation scheduling and loading,
- material staging and issue,
- priority control in work in progress (WIP),
- capacity control in WIP,
- quality control in WIP,
- manufacturing order close,
- process evaluation with regard to labour, material and equipment costs,
- facilitating the down-load of CAPP instructions.

Basically, PAC can be seen as the execution of the long term plans developed from the master production schedule and materials plan. It is a short term activity and, by nature, tends to be a *data intensive* activity.

The goals of PAC can be divided into four categories concerning the control of WIP, quality, labour and equipment, respectively. These goals are as follows.

(1) Work in progress
- reduced WIP investment,
- balanced workload,

- improved delivery performance,
- reduced manufacturing lead time.

(2) Quality
- reduced incidence of defects and scrap,
- reduced appraisal costs.

(3) Labour
- improved efficiency,
- improved utilization,
- increased operator satisfaction.

(4) Equipment
- improved utilization,
- improved availability,
- reduced set-up costs.

All of these goals are inextricably interrelated and are in some cases conflicting. For example, both WIP investment and manufacturing lead time could be reduced by sacrificing labour and machine utilization.

4.4 The role of PAC in CIM

In the previous sections we have placed PAC within the hierarchy of PMS and we shall now extend that description to cover its role in the CIM environment. This involves describing the PAC interfaces to some key islands of automation in the CIM environment. This is illustrated in Figure 4.6.

We shall now describe the following interfaces:

- automated material handling and storage systems,
- automation of the fabrication/assembly process,
- computer aided testing,
- computer aided process planning,

as well as the problems in linking PAC with other levels in PMS.

4.4.1 PAC and automated material handling and storage systems

The interface between PAC and Automated Material Handling and Storage Systems (AMHSS) is quite complex and information flows in both directions. PAC, as it were, supplies much of the intelligence to drive the material handling system. This intelligence is of three types:

(1) **Routing control** To which locations is a part to be moved?

(2) **Real-time material status** What is the current location of each part being moved?

(3) **Schedule control** When is a part to be routed to a location?

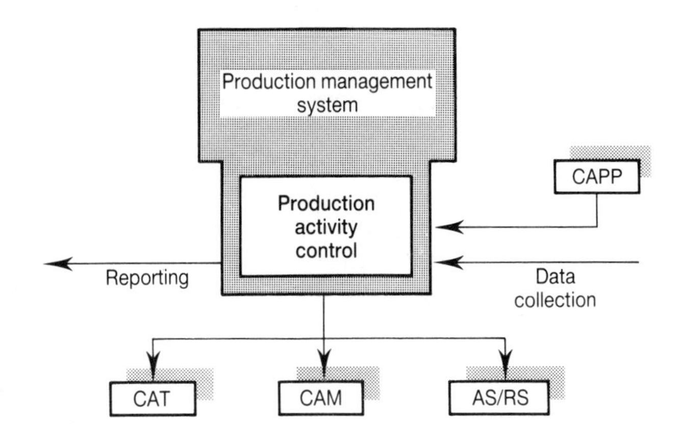

Figure 4.6 PAC interaction with CIM.

AMHSS has two functions:

(1) **Handling**, as evidenced by:
- Automatic Guided Vehicle Systems (AGVS),
- computer controlled conveyors,
- overhead delivery systems, and
- flexible transfer devices.

(2) **Storage**, as evidenced by:
- Automatic Storage and Retrieval Systems (AS/RS),
- automatic carousels, and
- automatic small parts storage and kitting systems.

It is at the interface between storage and handling (i.e. order picking) that the relationship to PAC is at its most complex. The bill of materials required for the order is compared with the shop routing to identify which work centre requires the material. Information on operation lead times is required to identify the lead time offset from order release until component/material delivery to the work centre. Intelligent PAC systems may have the ability to consolidate component issue requirements across both scheduled and planned orders in order to achieve picking efficiencies by using *optimal* issue quantities. Component issue schedules are presented in *optimal* location sequence so as to minimize retrieval times (this is a dynamic problem since storage locations are typically random). Finally, the picked material can be transferred automatically from the storage systems to the transport system. Current technology in PMS is typically very weak in optimizing the material issue logistics. Most systems typically offer issue to order functionality only.

Because the materials handling system is in such close contact with each stage of the production process, it provides a good foundation on which to build an automatic data collection system. The technology of automatic identification of bar-coded tote bins by laser scanners is already well established. The problem is in communicating this information to the PAC system. The linking of automatic data collection to the physical control of materials leads to almost 100% accuracy of WIP status.

One of the key PAC objectives is to minimize manufacturing lead time and WIP inventory. Too often systems designers look to the automation of materials handling to deliver this lead time reduction. Design for automated materials handling does place constraints on queue sizes and, consequently, lead time. However, it is not often understood that manufacturing lead time is primarily a function of the manufacturing process organization and that it is through combination or elimination of separate stages of the manufacturing process that lead time is primarily reduced. Setting constraints on inter-operation buffers is also important but automating the movement of material, in itself, frequently has little effect.

4.4.2 PAC and automation of fabrication/assembly

The relationship between PAC and process automation is much discussed in the case of existing and proposed flexible manufacturing systems, for example see Akella *et al.* (1984). The PAC system must schedule and manage communication between all the sub-systems of the FMS. In the future we will continue to see the proliferation of intelligent process equipment on the manufacturing floor. The PAC module will be the primary source of intelligence to drive this equipment and use will also be made of the local intelligence of this equipment to gather information from the factory floor. The principal examples of intelligent process equipment in the CIM environment will be as follows:

- robots,
- CNC/DNC systems and FMS cells,
- computer controlled processes,
- automatic and semi-automatic assembly equipment.

One principal role of the PAC system is to down-load process instructions, typically generated by Computer Aided Process Planning (CAPP). Production schedules may be communicated to robots and FMS cells. The PAC system automatically tracks tooling usage and controls automatic tool changes. The intelligent equipment relays information on operation completions and tool wear; it may also function as an automatic test equipment and report quality problems.

4.4.3 PAC and computer aided testing (CAT)

The nature of the link between CAT and PAC is similar to the link between PAC and the automation of fabrication and assembly. PAC down-loads test diagnostics to the CAT stations. PAC, as part of its routing control, manages the test and rework procedures for each part going through the process. Intelligent PAC systems can dynamically calculate *optimal* test and rework tactics based on analysis of historical information. The PAC system also feeds back corrective action requests from the CAT station to fabrication and assembly stations. In return for this, the CAT system can act as an automated data collection system, feeding information on quality events back to the PAC system in real time. The PAC system can direct exception messages to management when quality performance is unacceptable. The integration of PAC and CAT will most likely require the bar-coded identification of each individual unit, as opposed to each tote bin as would suffice in the case of stand alone PAC. The principal examples of CAT systems include:

- testing of physical dimensions after fabrication (perhaps by systems with vision capability),
- component verification as part of the assembly process,
- in-circuit testing after assembly,
- system functional testing after system configuration.

4.4.4 PAC and computer aided process planning

We have already discussed computer aided process planning. The problem of interfacing CAPP to a PAC system consists of communicating the routings, operation descriptions and process instructions (including NC part programs, tooling instructions and time standards) to the PAC database. The PAC system, because of its access to demonstrated historical performance on the factory floor, can relay and generate revisions to the parameters used in the CAPP process.

4.4.5 Interaction between PAC and other levels in PMS

The interaction of PAC to the higher levels in the PMS has been outlined above and is described in Harhen (1983) and Harhen and Browne (1984). The problem is that current PAC modules in production management systems lack the functionality to control and communicate with the factory floor. The key factors include:

- Absence of *off the shelf* interfaces to automatic data collection devices.
- Absence of quality management functionality.
- The routing control functionality of PAC modules in PMS is extremely naive. Only straight line flows are typically allowed and there is no support of multilevel tracking, i.e. tracking both the top level product and key sub-assemblies and components through parallel routings. In addition, tracking is primarily by manufacturing order and support is rarely given for individual identification of units within a batch.
- Tooling control is rarely adequately provided for.
- Preventative maintenance and equipment tracking is rarely available.
- Finally, in systems such as MRP II, relatively few systems provide for real-time schedule regeneration by bucketless net change MRP.

It is clear that there must be major enhancements to the PAC functionality of PMS before they can operate fully in the CIM environment. The need is for the development of PMS with control and communications capability sufficient for the CIM environment. These conclusions are supported by the CAM-I Factory Management Project (Computer Aided Manufacturing International 1983) comment on current MRP II systems:

> 'Although a large number of manufacturing control systems are currently on the market, the greatest emphasis is on bill of material processing and material requirements planning. A distributed system oriented for separate levels of factory management and which implements closed-loop communication and control, including the coordination of all shop service functions as well as material flow control, was not found'.

4.5 Conclusion

In this chapter we have examined the various levels of planning and control in manufacturing systems. In so doing, the different levels within the production management system hierarchy were briefly outlined. We then proceeded to discuss how the PMS is affected by different forms of manufacturing process organization and we noted that the production management problem may be more or less complex, depending on the manufacturing process organization within which it resides. We concentrated our discussion on production activity control, since PAC is the gateway between the execution layer on the factory floor and remainder of the manufacturing and production management system. We noted that, at the operational level, PAC is the module which integrates the various operational elements of the manufacturing system and ultimately leads to CIM on the shop floor. We went on to study the linkages of PAC to various other islands of automation within the CIM environment.

We will now go on in Parts II, III and IV to examine the underlying philosophies of the various approaches to production management, in order to understand the environment within which PAC exists, from a production management point of view.

References

Actis-Dato, M., Erhet, O. and Barta, G. 1986. 'Control systems for integrated manufacturing: the CAM solution', in *ESPRIT'85: Status report of ongoing work*, edited by the Commission of the European Communities. Amsterdam: North Holland.

Akella, R., Choong, Y. and Gershwin, S. 1984. 'Performance of hierarchical production scheduling policy', *IEEE Transactions on Computers, Hybrids and Manufacturing Technology* **CHMT-7** (3), September.

Anonymous 1981. 'Implementing CIM', *American Machinist*, August, 152–174.

Bachant, J. and McDermott, J. 1984. 'R1 revisited, four years in the trenches', *AI Magazine*, Fall, 21–32.

Baumol, W. and Braunstein, Y. 1977. 'Empirical study of scale economics and production complementarity; the case of journal productions' *Journal of Political Economy*, **85**(5), 1037–1048.

Bonsack, R. 1986. 'Cost accounting in the factory of the future' *CIM Review*, **2**(3) 28–32.

Boothroyd, G. and Dewhurst, P. 1983. *Design for Assembly: A Designers Handbook* Amherst, Mass: University of Massachusetts.

Bowden, R. and Browne, J. 1987. 'ROBEX – an artificial intelligence based process planning system for robotic assembly', in *Proceedings of the IXth ICPR*, edited by A. Mital. Cincinnati, USA: University of Cincinnati, College of Engineering.

Browne, J., Dubois, D., Rathmill, K., Sethi, S. and Stecke, K. 1984. 'Classification of flexible manufacturing systems', *The FMS Magazine*, April, 114–117.

Bullinger, H.J. and Ammer, E.D. 1984. 'Work structuring provides basis for improving organization of production systems', *Industrial Engineering*, **16**(10), 74–82.

Burbidge, J. 1986. 'Production planning and control: a personal philosophy', paper presented to IFIP Working Group 5.7 meeting in Munich, FRG, March.

Butcher, M. 1986. 'Advanced manufacturing system (AIMS) for aero engine turbine and compressor discs', in *Proceedings of the FMS5 Conference*, edited by K. Rathmill. UK: IFS Publications, 93–104.

Commission of the European Communities. 1982. *ESPRIT – The Pilot Phase*, COM (82) 486 Final 1/2, CEC. Brussels: Commission of the European Communities.

Computer Aided Manufacturing International. 1983. *CAM-I Factory Management Project, PR-82-ASPP-01.6*. USA: Computer Aided Manufacturing International Inc.

Conway, R., Maxwell, W. and Miller L. 1967. *Theory of Scheduling*, Reading, Mass: Addison-Wesley Publishing Co.

Cross, K.F. 1984. 'Production modules; a flexible approach to high tech manufacturing', *Industrial Engineering,* **16**(10), 64–72.

Descotte, Y. and Latombe, J.C. 1981. 'GARI: a problem solver that plans how to machine mechanical parts', *IJCAI*, 766–772.

Descotte, Y. and Latombe, J.C. 1985. 'Making compromises among antagonistic constraints in a planner, *Artificial Intelligence*, **27**, 183–217.

Forrester, J. 1961. *Industrial Dynamics*. Cambridge, Mass: MIT Press.

French, R. 1980. 'Accurate work in process inventory: a critical MRP system requirement', *Production and Inventory Management*, First Quarter, 17–22.

Gerwin, D. 1982. 'Do's and don'ts of computerized manufacturing', *Harvard Business Review*, **60**(2), 107–116.

Gerwin, D. and Tarondeau, J.C. 1986. 'Consequences of programmable automation for French and American automobile factories: an international case study', in *Production Management: Methods and Studies*, edited by B. Lev. Amsterdam: Elsevier Science Publishers, 85–98.

Gold, B. 1986. 'CIM dictates change in management practice', *CIM Review*, **2**(3), 3–6.

Goldhar, J. 1983. 'Plan for economies of scope', *Harvard Business Review*, **61**(6), 141–148.

Goldhar, J. and Jelinek, M. 1985. 'Computer integrated flexible manufacturing: organizational, economic and strategic implications', *Interfaces*, **15**(3) 94–105.

Gould, L. 1985. 'Computers run the factory', *Electronics Week*, March 25.

Groover, M. 1980. *Automation, Production Systems, and Computer Aided Manufacturing*, New Jersey, USA: Prentice-Hall Inc.

Gunn, T. 1982. 'The mechanization of design and manufacturing', *Scientific American*, **247**(3), 87–110.

Harhen, J. 1983. *Production Activity Control: Systems Design and Implementation*, M.Eng.Sc. thesis, University College Galway, Ireland.

Harhen, J., Browne, J., and O'Kelly M. 1983. 'Production activity control and the new way of life', *Production and Inventory Management*, Fourth Quarter, 73–85.

Harhen, J. and Browne, J. 1984. 'Production activity control; a key node in CIM', in *Strategies for Design and Economic Analysis of Computer Supported Production Management Systems*, edited by H. Hubner. Amsterdam: North Holland.

Harhen, J., Cohen, P., Graves, R. and Ketcham, M. 1987. 'Using multiple perspectives in manufacturing macro-planning', in CIM Europe 1987 Conference Proceedings. UK: IFS Publications.

Harley, M., Joyce, R., Cooney, B. and Browne, J. 1986a. 'A specification for a production activity control (PAC) system in a CIM environment', in *Proceedings of CAPE 1986 Conference*, edited by K. Bo, L. Estensen and E. Warman. Copenhagen: North Holland.

Harley, M., Joyce, R., Cooney, B. and Browne, J. 1986b. 'A SPECIF model of PAC for CIM', in *Proceedings of CIM Europe Conference, Bremen, FRG*, edited by B. Hirsch and M. Actis-Dato. Amsterdam: North-Holland.

Hayes, R. and Wheelright, S. 1984. *Restoring our Competitive Edge: Competing through Manufacturing*. New York: John Wiley and Sons.

Hitomi, K. 1979. *Manufacturing Systems Engineering*. London: Taylor and Francis.

King, J. 1976. 'The theory practice gap in job shop scheduling', *The Production Engineer*. March, 137–143.

Lyneis, J. 1980. *Corporate Planning and Policy Design: A Systems Dynamics Approach*. Cambridge, Mass: MIT Press.

McDermott, J. 1981. 'R1: the formative years', *AI Magazine*, Summer, 21–29.

Merchant, M. 1977. 'The inexorable push for automated production', *Production Engineering*, January, 44–49.

Meredith, J. and Suresh, N. 1986. 'Justification techniques for advanced manufacturing technologies', *International Journal of Production Research*, **24**(5), 1043–1057.

Nanda, R. 1986. 'Redesigning work systems – a new role for IE', in *Proceedings of the 1986 Fall Industrial Engineering Conference*, AIIE, 222–229.

Primrose, P. and Leonard, R. 1986. 'Conditions under which flexible manufacturing is financially viable', in *Flexible Manufacturing Systems: Methods and Studies*, edited by A. Kusiak. Amsterdam: North-Holland.

Roberts, E. 1978. *Managerial Applications of Systems Dynamics*. Cambridge Mass: MIT Press.

Rolstadas, A. 1986. 'Trends in production management systems', in *Advances in Production Management Systems 85*, edited by E. Szelke and J. Browne. Amsterdam: North Holland.

Rosenthal, S. and Ward, P. 1986. 'Key managerial roles in controlling progress towards CIM', in *Manufacturing Research: Organizational and Institutional Issues*, edited by A. Gerstenfeld, H. Bullinger, and H. Warnecke. Amsterdam: Elsevier Science Publishers.

Rygh, O. 1980. 'Succeeding with AS/RS technology in the 80s', *Industrial Engineering*, September, 56–63.

Sanderson, R.J., Campbell, J.A. and Meyer, J.D. 1982. *Industrial Robots, A Summary and Forecast for Manufacturing Managers*, Tech Tran Corporation, USA.

Shingo, S. 1981. *Study of Toyota Production System from Industrial Engineering Viewpoint*. Japanese Management Association, 352.

Skinner, W. 1969. 'Manufacturing – the missing link in corporate strategy', *Harvard Business Review*, May/June, 156.

Skinner, W. 1985. *Manufacturing: The Formidable Competitive Weapon*. New York: John Wiley and Sons.

Spur, G. 1984. 'Growth, crisis and the factory of the future', *Robotics and Computer Integrated Manufacturing*, **1**(1), 21–37.

Stecke, K. and Browne, J. 1985. 'Variations in flexible manufacturing systems according to the relevant types of automated materials handling', *Material Flow*, **2**, 179–185.

Talaysum, A., Hassan, M., Wisnosky, D. and Goldhar, J. 1986. 'Scale vs. scope: the long run economics of the factory of the future', in *Advances in Production Management Systems 85*, edited by E. Szelke and J. Browne. Amsterdam: North-Holland.

Westkamper, E. 1986. 'Increase in flexibility and productivity with computer integrated and automated manufacturing', in *Proceedings of FMS5*, edited by K. Rathmill. UK: IFS Publications, 121–126.

Weston, R., Sumpter, C. and Gascoigne, J. 1986. 'Distributed manufacturing systems', *Robotica*, **4**(1), 15–26.

Wild, R. 1971. *The Techniques of Production Management*. London: Holt, Rinehart and Winston.

Yeomans, R. 1985. *Design Rules for a CIM System*. Amsterdam: North-Holland.

PART II

The requirements planning approach: MRP and MRP II

Overview

Material Requirements Planning (MRP) and Manufacturing Resource Planning (MRP II) have, almost certainly, been the most widely implemented large scale production management systems since the early 1970s. Several thousand systems of this style are in use in industry around the world. The aim in Part II of this book is to describe and review the MRP approach. This discussion addresses both material requirements planning, which generates a schedule of manufacturing and purchase orders to meet a given demand, and manufacturing resource planning, which is an extension of MRP to support the integrated management of many of the functions of the manufacturing enterprise. An attempt will be made to give the reader an insight into the assumptions and techniques of the material requirements planning and manufacturing resource planning approach. Part II is organized in the following manner.

Chapter 5 serves as an introduction to MRP. It provides some insight into the history of MRP and identifies the important assumptions and attributes of the MRP approach. The approach is illuminated by the presentation of an MRP example.

In Chapter 6, the use of the more important techniques within the MRP system is discussed. This discussion covers such topics as bottom-up replanning using pegged requirements and the firm planned order. The net change versus regenerative planning approaches to requirements planning is discussed, as are bucketed and bucketless systems.

In Chapter 7, *closed loop* MRP is described, together with its evolution to the extended version of MRP known as MRP II. Master production schedule development, resource and capacity planning, and finally production activity control are also discussed.

Chapter 8 describes the manufacturing database while Chapter 9 presents a review of lot sizing approaches used within MRP.

In Chapter 10 the status of MRP/MRP II as a paradigm for production management is discussed. In so doing, the effectiveness for MRP II implementations and practice is covered. Various criticisms of the MRP approach are reviewed and some of the current trends in MRP research are described. Finally, some of the philosophical debate that surrounds MRP is examined. This chapter serves as the basis for a comparative review of alternative approaches presented in Part V.

It will be seen that MRP/MRP II is certainly a viable approach to production management with a proven track record. Although MRP II will continue to be widely applied in its present form, it will likely be subject to radical modularization in the future in order to re-emerge, in new hybrid production management environments that complement the other production management paradigms described in Parts III and IV.

CHAPTER FIVE

Introduction to requirements planning (MRP and MRP II)

5.1 Introduction

Material Requirements Planning (MRP) has been the most widely implemented large scale production management system since the early 1970s, with several thousand MRP type systems implemented in industry around the world. The aim in this chapter is to describe what requirements planning is. The history of requirements planning and the assumptions that underlie its application are discussed. The key attributes of the approach are described and its operation is illustrated by an example.

5.2 History of requirements planning

MRP originated in the early 1960s in the USA as a computerized approach for the planning of materials acquisition and production. The definitive textbook on the technique is by Orlicky (1975a). The technique had undoubtedly been manually practised in aggregate form prior to the second world war in several locations in Europe. However, what Orlicky realized was that a computer enabled the detailed application of the technique, which would make it effective in managing manufacturing inventories.

These early computerized applications of MRP were built around a Bill of Material Processor (BOMP) which converted a discrete plan of production for a parent item into a discrete plan of production or purchasing for component items. This was done by exploding the requirements for the top level product, through the Bill of Material (BOM), to generate component demand. The projected gross demand was then compared with available inventory and open orders over the planning time horizon and at each level in the BOM. These systems were implemented on large mainframe computers and run in centralized material departments for large companies.

As time passed, the installations of the technique became more widespread and various operational functions were added to extend the range of tasks that these software systems supported. In particular, these extensions included Master Production Scheduling (MPS), Production Activity Control (PAC), Rough Cut Capacity Planning (RCCP), Capacity Requirements Planning (CRP), and Purchasing.

The combination of the planning (MPS, MRP, CRP) and execution modules (PAC and purchasing) with the potential for feedback from the execution cycle to the planning cycle, was termed **closed loop MRP**. With the addition of certain financial modules, as well as the extension of master production scheduling to deal with the full range of tasks in master planning and the support of business planning in financial terms, it was realized that the resultant system offered an integrated approach to the management of manufacturing resources. This extended MRP was labelled **manufacturing resource planning** or **MRP II**. Since 1980, the number of MRP installations has continued to increase as MRP applications became available at lower cost on mini-computers and micro-computers.

While it is difficult to assess the total number of MRP implementations, it is reasonable to suggest that it is of the order of several thousand. The 1982 survey by Anderson *et al.* (1982) of APICS members in two of the 14 APICS regions in the USA indicated 433 companies using MRP in these two regions alone. Promotional material from the more popular independent software companies selling MRP packages often claims of the order of 200 to 600 implementations each. A recent study by LaForge and Sturr (1986) of firms employing 100 or more employees in South Carolina, USA, indicated that 31% of these firms were users of MRP. Wight (1981) presents a claim that 8000 firms in the USA were using some form of MRP by mid-1981.

MRP's popularity stems back to the *MRP crusade* launched by the American Production and Inventory Control Society (APICS) in the early 1970s. The focus of this was to convince people that MRP was the solution, since it represented an integrated communication and decision support system that supports the management of the total manufacturing business. It was emphasized that in order to succeed, MRP implementation programmes required management commitment and total workforce education. The role of optimizing techniques drawn from operations research and management science was frowned upon. The real problems, it was declared, were problems of discipline, education, understanding and communication. This message was promoted by APICS and a stream of almost evangelical consultants, and finally echoed by a computer industry eager to expand the range of applications it could offer.

One of the significant reasons MRP was adopted so readily as the production management technique was that it made use of the computer's ability to store centrally and provide access to the large body of information that seemed necessary to run a company. It helped to coordinate the activities of various functions in the manufacturing firm such as engineering, production and materials. Thus the attraction of MRP II lay not only in its role as decision making support, but, more importantly, in its integrative role within the manufacturing organization.

Today there is some concern as to how systems of the style of MRP can be integrated into a Computer Integrated Manufacturing (CIM) environment and the adequacy of such systems compared with alternate philosophies such as Kanban/Just In Time (JIT) and proprietary techniques such as Optimized Production Technology, (Fox 1985). Questions are being raised regarding the effectiveness of MRP, as managers see the extent of the effort required to implement such systems and the all too frequent failure to realize the promised benefits.

5.3 The attributes of material requirements planning

Prior to the widespread use of material requirements planning, the planning of manufacturing inventory and production was generally handled through inventory control approaches, for example:

- The **two bin policy** under which inventory availability is continuously reviewed and a predetermined quantity (fixed batch size) of items is ordered each time stocks fall below a predetermined level (reorder point).
- The **periodic order cycle policy** under which inventory is reviewed on a fixed periodic basis and sufficient items (a variable quantity) are ordered to bring the stock level up to a predefined level (target inventory).

The implicit assumption of these inventory control approaches is that the replenishment of inventory items can be planned independently of each other. The planning philosophy is that the inventory availability of each component should be maintained. Orlicky offered several important insights, which revolutionized manufacturing inventory management practice.

- Manufacturing inventory, unlike finished goods or service parts inventory, cannot usefully be treated as independent items. The demand for component items is dependent on the demand for the assemblies of which they are part.
- Once a time phased schedule of requirements for top level assemblies is put in place (master schedule), it follows that the dependent time-phased requirements for all components can be calculated. Consequently, it makes little sense to forecast them.
- The assumptions underlying inventory control models usually involve a uniform or at least a well defined demand pattern. However, the dependency of component demand on the demand for their parents gives rise to a

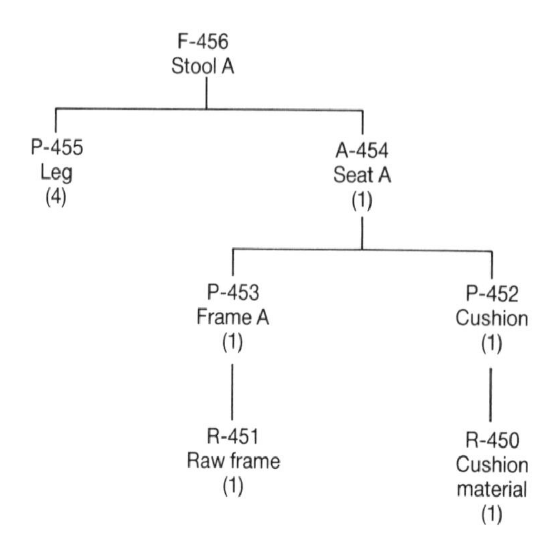

Figure 5.1 Product structure for Stool A.

phenomenon of discontinuous demand at the component level. Orlicky termed this *lumpy* demand. Lumpy demand occurs even if the master scheduled parts face uniform demand because of the effects of lot sizing and the fact that demand for an item often arises from a number of product sources. The implication of the lumpy demand phenomenon is that order point techniques are inappropriate for managing manufacturing inventories.

- A computer provides the data processing capability to perform the necessary calculations efficiently.

The shift from the stock control approach to the MRP approach, can be viewed as a shift from control of the level of stock to discrete control on the flow of material. MRP is a flow control system in the sense that it orders only what components are required to maintain manufacturing flow. Moreover, such orders can either be for purchased parts or manufactured parts. Therefore a requirements planning system lays the basis for both production scheduling and raw materials purchasing.

MRP is to be seen as a priority planning system in that it determines requirements, but it does not acknowledge all constraints that exist in the planning problem, particularly capacity. In the case of material constraints it points out the constraint violation but leaves the replanning to the user. In this way, MRP tells the users what must be done in order to meet the master schedule, as opposed to what can be done.

The starting point for MRP thus is the recognition that products to be manufactured or assembled can be represented by a bill of material. A bill of

Table 5.1 BOM for Stool A, Seat A, Frame A and Cushion.

STOOL A

Part number	Description	Quantity per	Make or buy
P-455	Leg	4	B
A-454	Seat A	1	M

SEAT A

Part number	Description	Quantity per	Make or buy
P-453	Frame A	1	M
P-452	Cushion	1	M

FRAME A

Part number	Description	Quantity per	Make or buy
R-451	Raw frame	1	B

CUSHION

Part number	Description	Quantity per	Make or buy
R-450	Cushion material	1	B

material describes the parent/child relationship between an assembly and its component parts or raw material. This is illustrated for an example stool in Figure 5.1 and shown in tabular form in Table 5.1. As can be seen, the bills of material may have an arbitrary number of levels and will typically have purchased items at the bottom level of each branch in the hierarchy. An implicit assumption is that there is an adequate part numbering system in the company to differentiate all parts and components at various stages of manufacturing where planning intervention may be required. The MRP system is based very simply on the fact that the BOM relationship allows one to derive the demand for component material based on the demand for the parent item. MRP was thus proposed as a technique for managing dependent component demand by transmitting the independent demand for top level products and spares through the component hierarchy, as represented by the BOM.

An MRP system is driven by the master production schedule, which records the independent demand for top level items. It is derived from evaluating forecasts, customer orders and distribution centre requirements. MRP uses this requirements information, together with information on product structure from the bill of materials file, current inventory status from

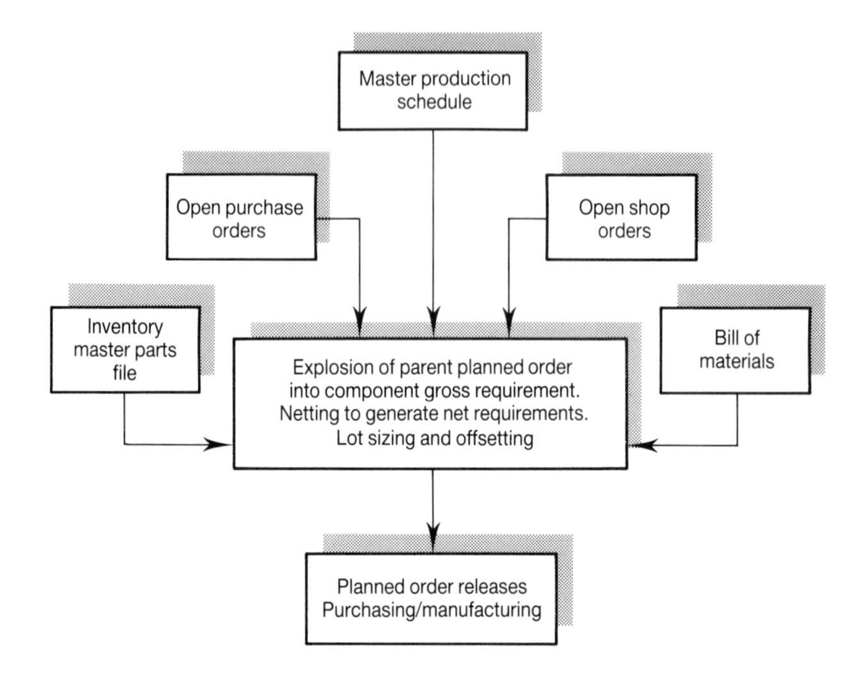

Figure 5.2 Basic structure of an MRP system.

the inventory file and component lead times data from the master parts file. MRP produces a time-phased schedule of planned order releases on lower level items for purchasing and manufacturing. This time-phased schedule is known as the materials requirements plan. This flow of information is illustrated in Figure 5.2.

Some of the main characteristics of MRP, which are apparent from the discussion above, include:

- MRP is *product oriented* in that it operates on a bill of materials to calculate the component and assembly requirements to manufacture and assemble a final product.
- MRP is *future oriented* in that it uses planning information from the master production schedule to calculate future component requirements instead of forecasts based on historical data.
- MRP involves *time phased requirements* in that during MRP processing, the requirements for individual components are calculated and offset by their expected lead time to determine the correct requirement date.
- MRP involves *priority planning* in that it establishes what needs to be done to meet the master schedule, as opposed to what can be done, given capacity and material constraints.

Table 5.2 Netting off MRP calculation.

	Gross requirements
+	Allocations
−	Projected inventory
−	Scheduled receipts
=	Net requirements

- MRP promotes control by *focusing on orders*, whether purchase orders or orders for the manufacturing plant.

5.4 How does MRP work?

In MRP, time is assumed to be discrete. Time is typically represented as a series of one week intervals, though systems which operate on daily planning periods are readily available. Demand for a component can derive from any of the products in which it is used, as well as independent spares parts orders.

A materials requirements planning system starts with the master production schedule as input and applies a set of procedures to generate a schedule of net requirements (and planned coverage of such requirements) for each component needed to implement the master production schedule. The system works down the BOM, level by level and component by component, until all parts are planned. It applies the following procedure for each component:

- Netting off the gross requirement against projected inventory and taking into account any open orders scheduled for receipt, as well as material already allocated from current inventory, thus yielding net requirements. The calculation is carried out as illustrated in Table 5.2.
- Conversion of the net requirement to a planned order quantity using a lot size.
- Placing a planned order in the appropriate period by backward scheduling from the required date by the lead time to fulfil the order for that component.
- Generating appropriate action and exception messages to guide the users attention.
- Explosion of parent item planned production to gross requirements for all components, using the bill of materials relationships.

As is apparent, the prerequisites to operate an MRP system include:

- A master production schedule must exist. This master production schedule is a clear statement of the requirements in terms of quantities and due dates for top level items.
- For every parent item there must be a corresponding bill of materials, which gives an accurate and complete statement of the structure of that item.
- For every planned part there must be a set of inventory status information available. Inventory status is a statement of physical stock on hand, material allocated to released orders but not yet drawn from physical stock and scheduled receipts for the item in question.
- For each planned part, either purchased or manufactured, a planning lead time must be set.

In order to explain these concepts more fully, the MRP technique will now be illustrated by a very simple example.

5.5 A simple MRP example

Consider the following example. Gizmo-Stools Inc. manufactures a simple four legged stool. The company has an order for 100 stools to be delivered four weeks from now. There are no stools or stool legs in stock. There are four legs per stool and the lead time to assemble the stools, once the legs are available, is one week. Consequently, 400 legs must be available three weeks from now. If we know that it takes approximately two weeks to manufacture the stool legs (i.e. the planning lead time for the stool legs is two weeks), then we must issue the order to manufacture the legs by the end of this week. The MRP calculations for the stool are illustrated in Table 5.3 and for the legs in Table 5.4.

Let us now extend this example to explore more fully the basic procedures of the MRP system. We will assume that Gizmo-Stools Inc. manufactures two types of stool, namely a four legged stool and a three legged stool. The product structures, in the form of product family tree diagrams, are shown in Figure 5.3.

Within the diagram we have indicated the quantity of each item per parent. Thus, for example, there are three legs per Stool B and four legs per Stool A. We also indicate the part number. Furthermore, we note that the two products have common components. Both stool types have the same legs, the same raw frames and the same cushion material. This situation is typical of the type of situation that MRP handles very well, i.e. where production covers a range of products with common components and sub-assemblies.

Table 5.3 Analysis for stool.

Item: Stool										
Week number	*1*	*2*	*3*	*4*	*5*	*6*	*7*	*8*	*9*	*10*
Gross requirements				100						
Scheduled receipts										
Projected inventory	0			−100						
Net requirements				100						
Planned orders			100							

Table 5.4 Analysis of stool leg.

Item: Stool leg										
Week number	*1*	*2*	*3*	*4*	*5*	*6*	*7*	*8*	*9*	*10*
Gross requirements			400							
Scheduled receipts										
Projected inventory	0		−400							
Net requirements			400							
Planned orders	400									

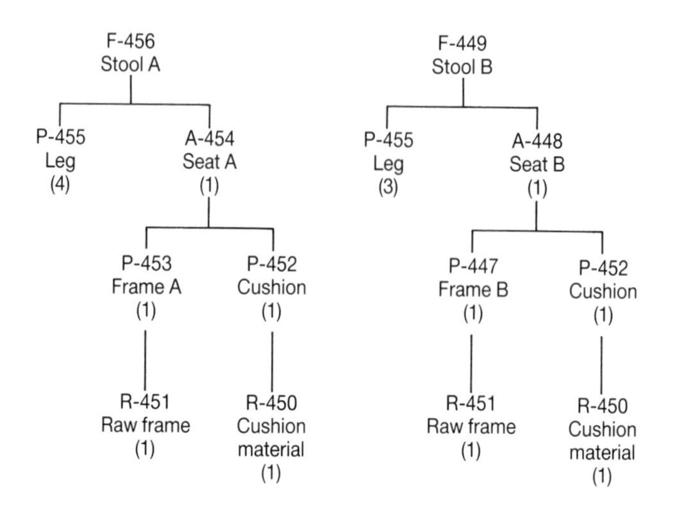

Figure 5.3 Two product structures.

In Table 5.5, the master parts information needed for the example is presented. The information is presented in part number order. The level code refers to the lowest level of any bill of material on which the component is to be found. Lead time is in weeks. The make/buy code refers to whether a part is manufactured or purchased. The lot sizing policy is **lot for lot** (L), by which the net requirement quantity is scheduled as the batch size for the replenishment order.

Within an MRP system, the **planning horizon** refers to the span of time the master production schedule covers, while the **time bucket** refers to the units of time into which the planning horizon is divided.

In this example a planning horizon of ten weeks and a time bucket of one week are used. In real MRP systems, time horizons should extend beyond the longest cumulative lead time for a product. Anderson *et al.* (1982) found in their study that the average MPS planning horizon was 40 weeks. The data structures used to represent time can be bucketed or non-bucketed. In the bucketed approach, a predetermined number of data cells are reserved to accumulate quantity information by period. In the non-bucketed approach, each part-quantity information pair has associated with it a time label. The bucketless approach is more flexible. The current week is beginning of week 1 and the master production schedule is shown in Table 5.6.

The relevant inventory data on each component are listed in Table 5.7. The **current inventory** represents the amount of material physically in stock. The **allocated** represents that quantity of physical stock already committed for released orders but not yet issued.

Table 5.5 Master parts data.

Level	Part number	Lot size	Lead time	Description	Make/Buy
0	F-449	L	2	Stool B	M
0	F-456	L	2	Stool A	M
1	A-448	L	1	Seat B	M
1	A-454	L	1	Seat A	M
1	P-455	L	2	Leg	B
2	P-447	L	1	Frame B	M
2	P-452	L	1	Cushion	M
2	P-453	L	1	Frame A	M
3	R-450	L	3	Cushion Material	B
3	R-451	L	2	Raw frame	B

Table 5.6 Master production schedule.

Week number	*1*	*2*	*3*	*4*	*5*	*6*	*7*	*8*	*9*	*10*
Stool A							50			80
Stool B						40			70	

Table 5.7 Inventory data.

Part number	Current inventory	Allocated
A-448	10	0
A-454	0	0
F-449	30	0
F-456	10	0
P-447	10	0
P-452	60	20
P-453	50	10
P-455	40	0
R-451	0	0
R-450	0	0

Table 5.8 Open orders data.

Part number	Scheduled receipts	Due date
A-454	40	2
P-455	20	3
R-450	100	1

Table 5.8 lists those orders which are open, i.e. due as scheduled receipts. The concept of allocated material is best explained by an example. For example, we see that there is a manufacturing order for Seat A (part number A-454) due for completion by the end of week 2. At this stage not all of the raw material to build this order has been drawn from the stockroom. There are quantities of the components of Seat A, i.e. Frame A (part number P-453) and the cushion (part number P-452), allocated to the released order for Seat A, and so this allocated material cannot be considered available to satisfy other material requirements.

There is now sufficient data on which to illustrate a simple MRP calculation. The analysis begins with the top level items in the bill of materials. We will start with Stool A (see Table 5.9). We take as our **gross requirements** for Stool A the requirements identified in the master production schedule. We are also aware that according to the inventory data there are 10 of Stool A in stock. There are no scheduled receipts for this item and hence the **net requirement** is as shown in Table 5.9. The **order release** date is calculated simply by offsetting the net requirement due date by the lead time. The analysis for the other item in the master production schedule, i.e. the Stool B, is similar and is presented in Table 5.10.

The next item in the bill of materials is the leg which is common to both types of stool. Initially we must generate the gross requirements for legs based on the planned orders for Stool A and Stool B. From our bill of materials data, we know that there are three legs per Stool B and four legs per Stool A. Hence the gross requirement for legs is as indicated in Table 5.11. There are also scheduled receipts for legs to be delivered in week 3 and these must be taken into account in our calculation of net requirements as in Table 5.11.

The calculation for Seat A, based on the planned orders for Stool A, is shown in Table 5.12. The calculation of the net requirements for Seat B, based on the planned orders for Stool B, is shown in Table 5.13.

Table 5.9 Analysis of Stool A.

Item: Stool A	**Part number: F-456**									
Week number	*1*	*2*	*3*	*4*	*5*	*6*	*7*	*8*	*9*	*10*
Gross requirements							50			80
Scheduled receipts										
Projected inventory	10						−40			−120
Net requirements							40			80
Planned orders					40			80		

Table 5.10 Analysis of Stool B.

Item: Stool B	**Part number: F-449**									
Week number	*1*	*2*	*3*	*4*	*5*	*6*	*7*	*8*	*9*	*10*
Gross requirements						40			70	
Scheduled receipts										
Projected inventory	30					−10			−80	
Net requirements						10			70	
Planned orders				10			70			

Table 5.11 Analysis of legs (common part).

Item: Leg	**Part number: P-455**									
Week number	*1*	*2*	*3*	*4*	*5*	*6*	*7*	*8*	*9*	*10*
Gross requirements				30	160		210	320		
Scheduled receipts			20							
Projected inventory	40		60	30	−130		−340	−660		
Net requirements					130		210	320		
Planned orders			130		210	320				

Table 5.12 Analysis of Seat A.

Item: Seat A	**Part number: A-454**									
Week number	*1*	*2*	*3*	*4*	*5*	*6*	*7*	*8*	*9*	*10*
Gross requirements					40			80		
Scheduled receipts		40								
Projected inventory	0	40			0			−80		
Net requirements								80		
Planned orders							80			

Table 5.13 Analysis of Seat B.

Item: Seat B	**Part number: A-448**									
Week number	*1*	*2*	*3*	*4*	*5*	*6*	*7*	*8*	*9*	*10*
Gross requirements				10			70			
Scheduled receipts										
Projected inventory	10			0			−70			
Net requirements							70			
Planned orders						70				

We now go on to consider the requirements for Frame A (see Table 5.14), which is based on the planned orders for Seat A calculated previously. Next, the requirements for Frame B (Table 5.15) are reviewed, based on the results of the analysis of the needs of Seat B analyzed earlier. The requirements for the cushion, given that it is common to both types of stool, is based on the net requirements for Seat A and Seat B. The calculation is shown in Table 5.16.

Similarly the requirement for the raw frame is based on both types of frame and is shown in Table 5.17. The requirement for the cushion material is based on the net requirement for the cushion itself and is calculated in Table 5.18.

Thus we have managed to make our way down through the two relevant BOMs. For each component we have taken into account the gross requirements and calculated an appropriate schedule of planned orders based on inventory data and planning lead times. Finally, we exploded the planned orders to gross requirements for all component parts. This concludes our treatment of the introductory MRP example.

Table 5.14 Analysis of Frame A.

Item: Frame A	**Part number: P-453**									
Week number	*1*	*2*	*3*	*4*	*5*	*6*	*7*	*8*	*9*	*10*
Gross requirements							80			
Scheduled receipts										
Projected inventory	40						−40			
Net requirements							40			
Planned orders						40				

Table 5.15 Analysis of Frame B.

Item: Frame B	**Part number: P-447**									
Week number	*1*	*2*	*3*	*4*	*5*	*6*	*7*	*8*	*9*	*10*
Gross requirements						70				
Scheduled receipts										
Projected inventory	10					−60				
Net requirements						60				
Planned orders					60					

Table 5.16 Analysis of cushion.

Item: Cushion	**Part number: P-452**									
Week number	*1*	*2*	*3*	*4*	*5*	*6*	*7*	*8*	*9*	*10*
Gross requirements						70	80			
Scheduled receipts										
Projected inventory	40					−30	−110			
Net requirements						30	80			
Planned orders					30	80				

Table 5.17 Analysis of raw frame.

Item: Raw frame	**Part number: R-451**									
Week number	*1*	*2*	*3*	*4*	*5*	*6*	*7*	*8*	*9*	*10*
Gross requirements					60	40				
Scheduled receipts										
Projected inventory	0				−60	−100				
Net requirements					60	40				
Planned orders			60	40						

Table 5.18 Analysis of cushion material.

Item: Cushion material				**Part number: R-450**						
Week number	*1*	*2*	*3*	*4*	*5*	*6*	*7*	*8*	*9*	*10*
Gross requirements					30	80				
Scheduled receipts	100									
Projected inventory	100				70	−10				
Net requirements						10				
Planned orders			10							

5.6 Conclusion

In this chapter the MRP approach was introduced. Its assumptions were described and its operation illustrated by means of a simple example. Much of what remains to be known about the mechanics of the technique of MRP is really implementation detail, which is important from the perspective of operating an MRP system. Some factors fall within the domain of software engineering, some within the domain of decision science. None really change the fundamental procedure. In any MRP system, the decision making procedures are no more complex than the basic arithmetic illustrated in this chapter. These implementation factors are discussed mainly in Chapter 6. Extensions of MRP are described in Chapter 7. Discussion of the manufacturing database is left until Chapter 8 and lot sizing techniques until Chapter 9.

CHAPTER SIX

The use of the MRP system

6.1 Introduction

An overview of the MRP approach to production planning was presented in Chapter 5 through a simple example. In this chapter, several important aspects of the operation of an MRP system are presented. The issues covered include such concerns as:

- Top-down planning in MRP.
- Bottom-up replanning
- Time representation in MRP systems.
- The role of safety stocks in an MRP system.

Each of these issues will now be discussed in turn.

6.2 Top-down planning with MRP

Change is continuous within the manufacturing environment. The master schedule changes. The inventory status changes. Engineering activity modifies BOMs. Orders are released to the shop floor or purchasing. Orders are completed. Some of these events are planned. Some are unplanned. For example, if an order is completed on time then this is a planned event. In this case the original material plan should still be valid. If, however, an order is completed ahead of time or is late, then this usually means that the material plan is no longer valid. In either case, the MRP system must accommodate such changes. There are two basic approaches to replanning within MRP systems. These are top-down planning and bottom-up replanning. Section 6.2 discusses top-down planning and bottom-up replanning is covered in Section 6.3.

We will discuss four important aspects of top-down planning:

(1) Regenerative planning and net change.
(2) The frequency of top-down planning.
(3) The use of low level coding.
(4) Rescheduling in top-down planning.

6.2.1 Regenerative and net change MRP

There are two basic styles to top-down planning. These are termed the **regenerative** approach and the **net change** approach. These involve alternative approaches to the system-driven recalculation of an existing material plan based on changes in the input to that plan.

Regenerative MRP starts with the master production schedule and totally re-explodes it down through all the bills of materials to generate valid priorities. Net requirements and planned orders are completely *regenerated* at that time. The regenerative approach thus involves a complete re-analysis of each and every item identified in the master schedule, the explosion of all relevant BOMs and the calculation of gross and net requirements for planned items. The entire process is carried out in a batch processing mode on the computer and, for all but the simplest of master schedules, involves extensive data processing. Because of this, regenerative systems are typically operated in weekly and occasionally monthly replanning cycles.

In the net change MRP approach, the materials requirements plan is continuously stored in the computer. Whenever there is an unplanned event, such as a new order in the master schedule, an order being completed late or early, scrap or loss of inventory or engineering changes to the BOMs, a partial explosion is initiated only for those parts affected by the change. If an event is planned, for example when an order is completed on time, then the original material plan should still be valid. The system is updated to reflect the new status but replanning is not initiated. Net change MRP can operate in two ways. One mode is to have an on-line net change system by which the system reacts instantaneously to unplanned changes as they occur. In most cases, however, change transactions are batched (typically by day) and replanning happens over night.

In the regenerative approach there is a vulnerability because of the need to maintain the validity of the requirements plan between system driven replanning runs. The role of user driven bottom-up replanning is discussed in Section 6.3. The MRP system also supports transactions that modify the status of the various planning inputs, such as inventory status, the MPS or the BOMs. However, in a regenerative system, these changes are only reflected in a new requirements plan after a new planning run. Net change systems typically operate with many frequent partial replanning

Table 6.1 Original master production schedule.

Week number	*1*	*2*	*3*	*4*	*5*	*6*	*7*	*8*	*9*	*10*
Stool A							50			80
Stool B						40			70	

Table 6.2 Revised master production schedule.

Week number	*1*	*2*	*3*	*4*	*5*	*6*	*7*	*8*	*9*	*10*
Stool A							50	10		80
Stool B						40			70	

runs and, as a result, are not subject to the same degree of vulnerability in plan validity between runs.

The difference between regenerative and net change can be illustrated by considering how each system treats the master schedule in replanning. Regenerative systems view the master schedule as a document, new editions of which are released on a periodic basis. Net change systems, on the other hand, see the master schedule as a document in a state of continuous change. The master schedule is processed in terms of the changes which have taken place since the last run. If we recall our simple MRP example from Chapter 5, we can illustrate this difference. The original master schedule is as shown in Table 6.1.

In our example in Chapter 5, we calculated the net requirements and planned order releases to meet this master schedule. Now consider what happens if, at the beginning of week 2, a new order comes in from a customer for 10 of Stool A to be delivered in week 8. This order is placed on the master schedule which now appears as illustrated in Table 6.2.

In a regenerative MRP approach, this new master schedule would be the basis for a *complete rerun* of the MRP analysis at the next replanning cycle. The analysis would proceed in the same manner adopted in Chapter 5. In a net change approach, it is the *change* to the master schedule that forms the basis of replanning, as in Table 6.3.

Table 6.3 Net change in the master production schedule.

Week number	*1*	*2*	*3*	*4*	*5*	*6*	*7*	*8*	*9*	*10*
Stool A								+10		
Stool B										

Under net change, it is only necessary to replan the requirements for Stool A. This replanning is executed by exploding the change in the master schedule through the bill of materials for Stool A and modifying all requirements information from the previous analysis. Parts unique to Stool B, such as Frame B (part number P-447) would not be replanned, whereas in a regenerative mode they would have been. We note that a net change system must store all the requirements analysis on an ongoing basis, whereas in a regenerative system this is not necessary, but may be done.

One potential difficulty with net change systems is that there is a reduced self purging capability. Errors may creep into the requirements plan, perhaps because of planner actions in bottom-up replanning. Since the master schedule is not completely re-exploded, as in the regenerative approach, any errors in the old plan tend to remain. As a result, errors in the material plan may accumulate over time. In order to counteract this problem, firms using net change MRP tend occasionally to do a complete regeneration so as to purge the system of these errors.

The survey of 433 companies using MRP conducted by Anderson *et al.* (1982) indicated that 30.3% of those studied were using the net change approach. The later study by LaForge and Sturr (1986) found that 38% were using net change. Wemmerlov's in-depth survey (1979) of 13 MRP installations indicated that 5 of the 13 were using net change MRP and these tended to be the larger companies, using their own software, with the most experience of MRP and producing the most complex products.

The net change approach offers the advantages of reactiveness. The pressures for competitiveness will tend to encourage the migration towards net change style systems. The disadvantage of a net change system is its *nervousness*. A badly run net change system, as it reacts to many unplanned events, will flood the materials planners with exception messages.

6.2.2 Frequency of replanning

As we have seen, regenerative systems are typically replanned on a weekly or monthly basis. Net change systems support more frequent replanning, either on-line or batched in daily or weekly increments. There is a tradeoff between data processing costs and the maintenance of valid priorities on manufacturing and purchase orders. The consensus view seems to be that the replanning cycle should be no longer than a week.

Another insight into practice can be gained from knowing how often individual users update their master schedules. The survey conducted by Anderson *et al.* (1982) found that 56.7% of MRP users updated their MPS on a weekly replanning cycle, while 16.37% updated the MPS on a daily basis. The LaForge and Sturr study (1986) found that 45% were updating their MPS on a weekly basis, while 24% were updating daily.

6.2.3 Low level codes

Low level codes determine the sequence in which the processing of part requirements is carried out. Components may be common to many bills of materials. In regenerative systems, if MRP processing were simply to follow a path through the bill of material hierarchies in its replanning, it would, as a result, replan common components several times over. Low level coding is a data processing mechanism which serves to overcome this inefficiency. The procedure is to assign to each component a code which designates the lowest level in any bill of material on which it is found. MRP processing then can proceed level by level and a component will not be planned until the level currently being processed is that of its low level code. Low level codes are also a useful feature in net change systems when these systems are operated on a batch basis.

6.2.4 Rescheduling in top-down planning

Ho *et al.* (1986) describe rescheduling as one of the difficult problems to resolve in production scheduling, within an MRP context. When an MRP system is replanning in a top-down fashion, it typically will adjust either the due date or the quantity of any planned order. If it identifies the need to make a change to an open order (a scheduled receipt), it typically sends an exception message for the materials planner to execute the change.

Consider the following situation. The requirements for Stool A, as determined from calculation based on the master schedule at week 1, are shown in Table 6.4. A fixed lot size of 200 is being used. (Lot sizing techniques will be discussed in more detail in Chapter 9.) Analysis based on this master schedule results in the planned release of an order for 200 of Stool A in week 5.

At the beginning of week 2 a new master schedule is made available, which results in new gross requirements for Stool A. The new master schedule involves an extra requirement for 70 of Stool A in week 4. If we analyze the effects of this change we see that we can meet the overall requirement, including the new requirement for 70 in week 4, by rescheduling the original planned order for 200 from week 5 into week 2. (See Table 6.5.)

Table 6.4 Analysis of Stool A.

Item: Stool A	**Part number: F-456**									
Week number	*1*	*2*	*3*	*4*	*5*	*6*	*7*	*8*	*9*	*10*
Gross requirements							50			80
Scheduled receipts										
Projected inventory	10						−40			−120
Net requirements							40			80
Order coverage							200			
Planned orders					200					

Table 6.5 Analysis of Stool A based on a revised master schedule.

Item: Stool A	**Part number: F-456**									
Week number	*2*	*3*	*4*	*5*	*6*	*7*	*8*	*9*	*10*	*11*
Gross requirements			70			50			80	
Scheduled receipts										
Projected inventory	10		−60			−110			−190	
Net requirements			60			50			80	
Order coverage			200							
Planned orders	200 ←									

The type of situation just described occurs frequently in manufacturing. We may need to reschedule existing planned orders because of modifications to the master schedule, or because of failure of a vendor or shop to deliver in the planned lead time or, indeed, any unplanned event. Rescheduling may involve the retiming of a planned order, as in the example above, or it may require the modification of the order size or perhaps both.

The example above was a trivial one and the solution presented itself readily. However, how would we have coped if the new item in the master schedule could not be delivered within the available lead time? Let us review Table 6.6.

This revision of the gross requirements for Stool A shows a requirement for 70 of Stool A during week 3. This produces a net requirement of 60 in week 3. However the lead time for this item is 2 weeks which implies that the order should be scheduled for week 1 and we are presently in week 2.

Orlicky (1976) suggested the use of a **minimum lead time** to deliver such an order (purchasing or manufactured). This minimum lead time is the time required to complete an order under the highest priority and in the case of a manufactured item might be little greater than the sum of the processing time and the transportation times. The result of this is that we could define two lead times and use the *minimum lead time* for automatic rescheduling in the event of a problem such as that in Table 6.6.

MRP practice did not follow Orlicky's suggestion in this case. If faced with a situation like the above, the system will send an exception message to the planner and leave it to him/her to resolve the problem by the use of bottom-up replanning.

6.3 Bottom-up replanning

As pointed out earlier, an MRP system must react to change. In top-down planning, the system itself does the planning. An alternative is for the planner to manage the replanning process. This is termed bottom-up replanning and makes use of two tools – the pegged requirements report and the firm planned order both of which are described in this section.

6.3.1 Pegged requirements

Pegging allows the user to identify the sources of demand for a particular component's gross requirements. These gross requirements typically originate either from its parent assemblies or from independent demand in the MPS, or from the demand for spare parts. This process is again illustrated by recourse to our simple MRP example.

Table 6.6 Analysis of Stool A based on a further revised master schedule.

Item: Stool A	**Part number: F-456**									
Week number	*2*	*3*	*4*	*5*	*6*	*7*	*8*	*9*	*10*	*11*
Gross requirements		70				50			80	
Scheduled receipts										
Projected inventory	10	−60				−110			−190	
Net requirements		60				50			80	
Order coverage						200				
Planned orders	???			200						

Table 6.7 Analysis of legs (common part).

Item: Leg	**Part number: P-455**									
Week number	*1*	*2*	*3*	*4*	*5*	*6*	*7*	*8*	*9*	*10*
Gross requirements				30	160		210	320		
Scheduled receipts			20							
Projected inventory	40		60	30	−130		−340	−660		
Net requirements					130		210	320		
Planned orders			130		210	320				

Table 6.8 Pegged requirements for the leg.

Requirement		**Source**	
Component quantity	*Week number*	*Parent*	*Parent quantity*
30	4	Stool B	10
160	5	Stool A	40
210	7	Stool B	70
320	8	Stool A	80

Within the BOMs for the two types of stool in our master schedule the leg is a common part. The gross requirements for the leg shown in Table 6.7 arise from a number of sources, as illustrated in Table 6.8.

A report, such as Table 6.8, allows the materials planner to *retrace* the steps of the MRP analysis and to understand the sources of the total gross requirements for the item in question. The procedure of identifying each gross requirement with its source at the next immediate higher level in the BOM is termed **single level** pegging. Through a series of single level pegging reports, we can eventually trace a set of requirements back to their sources in the master schedule.

An alternative facility is **full pegging** where each individual requirement for a planned item is identified against a master production scheduled item and/or a customer order. This, however, is quite rare in practice for a number of reasons. As is clear even from our simple example, requirements tend to arise from a number of sources in the master schedule. If a lot sizing technique is used, it becomes practically impossible to associate individual batches or lots with particular customer orders. Other factors, such as safety stocks, shrinkage and scrap allowances, further complicate the situation. As a result, the single level pegging facility is standard practice and full pegging is rarely used.

The technique of pegging is useful in that it allows the user to retrace the MRP systems planning steps in the event of an unexpected event, such as a supplier being unable to deliver in the planning lead time. By retracing the original calculations the user can detect what orders are likely to be affected and perhaps identify appropriate remedial action. The remedial action often involves overriding the normal planning procedures of MRP. This is done by using the **firm planned order** technique.

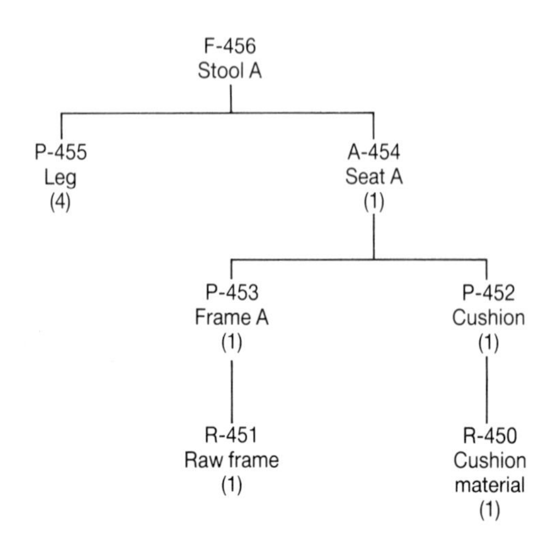

Figure 6.1 Product structure for Stool A.

6.3.2 Firm planned orders

The firm planned order allows the materials planner to force the MRP system to plan in a particular way, thus overriding lot size or lead time rules. A firm planned order differs from an ordinary planned order in that the MRP explosion procedure will not change it in any way. This technique can aid planners working with MRP systems to respond to specific material and capacity problems.

A typical problem might be the failure of a vendor or the manufacturing plant to deliver an order within the allocated lead time. Consider the BOM for Stool A in Figure 6.1 drawn from our earlier example.

It is clear from this BOM that Stool A cannot be produced unless Seat A and the legs are available. The lead time for the legs is two weeks, while that for Seat A is one week. The lead time for Stool A, assuming that Seat A and the legs are available, is two weeks. The lead time for Frame A and the cushion is one week.

Assume we have a master schedule requirement for 100 of Stool A in week 10. This leads us to plan the availability of 100 of Seat A by the end of week 8, and 100 each of Frame A and the cushion by the end of week 7. We also plan the availability of 400 of the legs at the end of week 8. The unexpected occurs! We are informed by the factory department responsible for the manufacture of Seat A that it cannot deliver the required amount of Seat A until week 9 – one week late.

What are the consequences of this?

- The delivery of the 100 items of Stool A is now in danger unless the lead time for assembling the stool can be reduced from two weeks to the one week available.
- If a **rush** job on the stool assembly is possible, then it is clear that there is now no necessity to have 400 of the legs available at the end of week 8. They can wait until the end of week 9 when Seat A will also become available.

In this type of situation, we need a capability to override the normal *logic* of MRP in order to cater for the unexpected event. Assuming that Stool A can be assembled by a *rush job* in one week, rather than the normal two weeks, there is no need for a change in the master schedule. We can create a *firm planned order* for Stool A for release by the end of week 9, rather than week 8, and mark it due for delivery by the end of week 10. The MRP explosion will, as a result, modify in the appropriate manner, the due dates for the legs, for Seat A and for all components of Seat A. The significance of the firm planned orders is that a subsequent MRP run will not attempt to *correct* the deviation from the normal planned lead times on Stool A.

The firm planned order can also override lot sizing rules (see Chapter 9). The availability of a *pegging* facility, combined with the use of firm planned orders, are the planner's chief tools in handling inevitable, unforeseen events in manufacturing.

6.4 Time representation

In this section **bucketed** and **bucketless** MRP systems will be discussed. Bucketed systems limit the time horizon that may be considered and the granularity of timing that may be ascribed to an order. Bucketless systems enable daily visibility to an order's date of requirement. The length of the planning horizon necessary to make an MRP system effective will also be discussed.

6.4.1 Bucketed and bucketless MRP systems

The data structures used to represent time in an MRP system can be bucketed or non-bucketed.

In the bucketed approach, a predetermined number of data cells are reserved to accumulate quantity information by period. This is illustrated by the matrix structure used in our calculations of requirements. These data cells are known as time buckets. A weekly time bucket contains all of the relevant planning data for an entire week. Since the number of buckets is

Table 6.9 Planned orders for legs.

Item: Leg	**Part number: P-455**									
Week number	*1*	*2*	*3*	*4*	*5*	*6*	*7*	*8*	*9*	*10*
⋮			⋮		⋮	⋮				
Planned orders			130		210	320				

predetermined, this means that there is a bound on the planning horizon, depending on what time divisions the buckets represent.

Weekly time buckets are considered to be the granularity necessary for near and medium term planning by MRP, whereas monthly buckets are considered too coarse. However, the normal bucket of one week may itself be too coarse to facilitate detailed short term planning. Further out in the planning horizon, monthly or perhaps quarterly time buckets are acceptable. There is no reason why the MRP system cannot accommodate a variable time bucket size over the span of the planning horizon.

There is, however, a complication arising from the notion of time buckets which is best illustrated by means of an example. Recall the MRP example, in which we generated requirements for the stool leg (part number P-455). These were in the form of planned orders and are reproduced in Table 6.9.

These orders are scheduled for release in weeks 3, 5 and 6. But what do we mean by an order scheduled for release in week 3? Do we intend that the order be released at the beginning of the week, in the middle of the week or at the end of the week? In effect, each of these orders is an event which must be scheduled for a point in time. Our time bucket represents a span of time. The problem is resolved by the MRP system designer and users adopting a convention. One may choose to adopt the last day within the time bucket as the order receipt date. The important point is to adhere to the convention once it has been adopted.

In the non-bucketed approach, each element of time phased data has a specific time label associated with it and is not accumulated into buckets. What this means is that there is provision for daily visibility on requirements timing. Moreover, the bucketless approach has no limit on the extent of the planning horizon. Consequently, the bucketless approach is more flexible.

The survey of Anderson *et al.* (1982) suggests that the vast majority (70.4%) of MRP users work in time buckets of one week.

6.4.2 The planning horizon

The planning horizon refers to the span of time from the current date out to some future date, over which material plans are generated. The chief factor in determining the planning horizon is the longest cumulative manufacturing and procurement time for a master scheduled item. If our horizon does not extend this far, then a new order at the end of the planning horizon may require a release of a purchase order last week, and the master schedule will be infeasible before we start!

A planning horizon longer than the cumulative product lead time will avoid this problem and will give sufficient visibility to facilitate material procurement. The planning horizon is often extended further than the longest cumulative lead time for the purpose of gaining visibility of manufacturing capacity needs in the future.

The upper limit on the range of the planning horizon is determined by our ability to make meaningful statements about the nature and contents of the master production schedule. The longer the planning horizon, the more difficult it is to make useful forecasts about the marketplace and the likely demand for products and end level items. The need to put in place a master schedule over this planning horizon is the chief vulnerability of MRP systems. As Burbidge (1985b) says, 'it is not given to man to tell the future'. A naïve reliance on a dubious MPS is a recipe for failure. In general, firms have not recognized the critical need for reducing procurement and manufacturing lead times as a means of improving MPS accuracy.

Anderson *et al.* (1982) found in their survey that the average length of the planning horizon used in MRP systems was of the order of 40 weeks.

6.5 The use of safety stocks

Safety stocks are a quantity of stock planned to be maintained in inventory, to protect against unexpected fluctuations in demand and/or supply. In this sense, safety stocks can be considered as a type of insurance policy to cover unexpected events, whether such events be the failure of a vendor to meet a promised delivery date or an unexpected increase in demand for the product. However, given the high cost of tying up capital in inventory, the use of safety stocks can be expensive.

Safety stocks can be incorporated into the MRP analysis. In the master parts database there is normally a field within each record which can indicate the safety stock of each part. Going back to our planning example of stool legs, let us assume that there is a safety stock requirement of 20. This implies that we should have at least 20 legs available at all times. The effect of this on the planned orders is illustrated in Table 6.10.

Table 6.10 Analysis with safety stock of legs (method A).

Item: Leg	**Part number: P-455**				**Safety stock: 20**					
Week number	*1*	*2*	*3*	*4*	*5*	*6*	*7*	*8*	*9*	*10*
Gross requirements				30	160		210	320		
Scheduled receipts			20							
Projected inventory	40	40	60	30	−130		−340	−660		
Net requirements					150		210	320		
Planned orders			150		210	320				

We could present Table 6.10 in an alternative manner by subtracting the safety stocks from the initial inventory and then calculating the item net requirements in the usual way. This is illustrated in Table 6.11. In the **projected inventory** row, we note that the value for week 1 has been reduced from 40 to 20. In this situation the meaning of the term projected inventory has, in effect, changed. Whereas in previous tables of this type the projected inventory referred to the quantity of material projected to be physically in stock less any allocations against open orders, it now refers to the projected physical stock less allocations, less safety stock.

6.5.1 Should safety stocks be used?

It is clear that if we insert a safety stock requirement for a high level item then we will automatically generate additional requirements for lower level dependent items. This distortion of demand may more than outweigh the benefit of safety stocks. Orlicky agreed that safety stocks should be used for independent demand items as protection primarily against forecast errors. He also granted that in the case of an unreliable supplier, safety stock (or safety lead time) could be used for the component in question.

DeBodt and Van Wassenhove (1983) in a case study, commented as follows:

Table 6.11 Analysis with safety stocks (method B).

Item: Leg	**Part number: P-455**				**Safety stock: 0**					
Week number	*1*	*2*	*3*	*4*	*5*	*6*	*7*	*8*	*9*	*10*
Gross requirements				30	160		210	320		
Scheduled receipts			20							
Projected inventory	20		40	10	−150		−360	−680		
Net requirements					150		210	320		
Planned orders			150		210	320				

> 'From our analysis . . . we concluded that, no matter what forecasting technique is being used, forecast errors are considerable . . . It follows that, in order to maintain a good customer service level, it will be necessary to have some safety provision on the finished product level. However, it is also necessary to protect the production departments against stockouts. Therefore, it may also be necessary to have safety stocks on (some) component levels . . .'

Besides the inventory cost, safety stocks have an additional disadvantage in the case of dependent demand items, since their use tends to cover up and make it easy to live with problems. The safety stocks may encourage the ongoing acceptance of a poor quality manufacturing process or poor delivery service by a supplier. As such, no one will be motivated to eliminate the problem. This discussion will be taken further when we consider the JIT (Just in Time) approach to production management in Part III. In any case, the use of safety stock across the board is unjustified.

In Wemmerlov's (1979) survey of 13 MRP installations in the USA, three companies were using safety stocks at all levels, five used them only on low level items and five companies applied safety stocks strictly to the end item or finished goods level.

6.5.2 How is the level of safety stock calculated?

The setting of safety stock quantities for dependent demand items in MRP tends, in practice, to be done on an *ad hoc* basis. For a particular item, it may simply reflect the *best guess* of the materials planner having regard to local knowledge of a particular supplier's past performance or even, perhaps, a whole industrial sector's characteristics.

The level of safety stock may alternately be calculated by reviewing the historical average demand for the item in question and then deciding to have safety stock to cover a given percentage of the likely demand over the lead time of the item. Let us illustrate this point from our ongoing example. Demand for legs (part number P-455) is 720 over a 10 week period, which averages 72 per week. The lead time for this item is 2 weeks. One policy might be to cover 25% of the lead time, i.e. maintain one half of a week's demand in stock as a reserve or safety stock. This suggests a safety stock of 36. A more conservative and, of course, expensive policy might be to cover 50% of the lead time and hold one week's stock or 72 items as safety stock. A similar margin of safety could also be achieved by the use of a safety lead time, whereby an additional safety lead time is added to the planning lead time to allow for unexpected events.

An open area of manufacturing research concerns the use of safety stocks for long lead time items. This can have the effect of improving a firm's ability to react quickly to unexpected demand. It is a matter of debate as to whether this is better handled by a floating MPS **hedge** or by using safety stock on a few key long lead time items.

6.6 Conclusion

In this chapter the operation of the MRP system was covered. Approaches for top-down and bottom-up replanning were described. The use of bucketless and bucketed systems, the length of the planning horizon and the use of safety stocks were also discussed. In Chapter 7 the extended version of MRP, known as manufacturing resource planning or MRP II, will be reviewed.

CHAPTER SEVEN

Manufacturing resource planning (MRP II)

7.1 Introduction

This chapter describes Manufacturing Resource Planning (MRP II). Manufacturing resource planning represents an extension of the features of the MRP system to support many other manufacturing functions beyond material planning, inventory control and BOM control. We shall review how the evolution from MRP to MRP II took place and examine the notion of *closed loop* MRP. Some of the major MRP II modules will also be outlined.

7.2 The evolution from MRP to MRP II

Manufacturing resource planning evolved from MRP by a gradual series of extensions to MRP system functionality. These extensions were natural and not very complicated as, for example, in the addition of transaction processing software to support the purchasing, inventory and financial functions of the firm. In supporting the extension of decision support, similar and quite reasonable assumptions are made and similar procedures to those of MRP are applied. In this way, MRP was extended to support Master Planning, Rough Cut Capacity Planning (RCCP), Capacity Requirements Planning (CRP) and Production Activity Control (PAC). Production activity control is the term favoured by the American Production and Inventory Control Society to cover activities traditionally described by shop floor control.

The term *closed loop MRP* denotes a stage of MRP system development wherein the planning functions of master scheduling, MRP and capacity requirements planning are linked with the execution functions of production

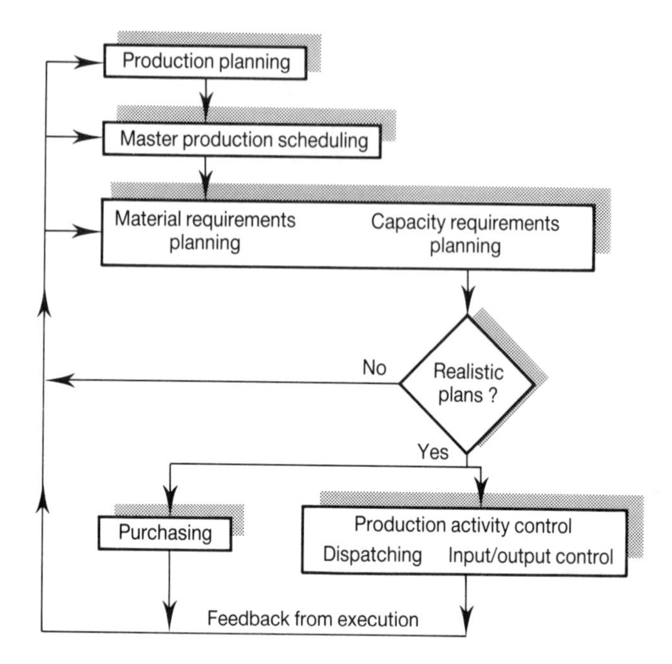

Figure 7.1 Closed loop MRP.

activity control and purchasing. These execution modules include features for input-output measurement, detailed scheduling and dispatching on the shop floor, planned delay reports from both the shop floor and vendors, as well as purchasing follow-up and control functionality. Closed loop signifies that not only are the execution modules part of the overall system, but there is also feedback from the execution functions so that plans can be kept valid at all times. Figure 7.1 indicates the nature of closed loop MRP.

With the extension of master production scheduling to deal with all master planning and the support of business planning in financial terms, and through the addition of certain financial features to the closed loop system so that outputs, such as the purchase commitment report, shipping budget and inventory projection could be produced, it was realized that the resultant system offered an integrated approach to the management of all manufacturing resources. This extended MRP was labelled *manufacturing resource planning* or *MRP II*. The MRP II system is thus a closed loop MRP system with additional features to cover business and financial planning. MRP II nominally includes an extensive **what if** capability. However, several of the additional features included in MRP II software packages remain unused in practice. We shall see why in Chapter 10.

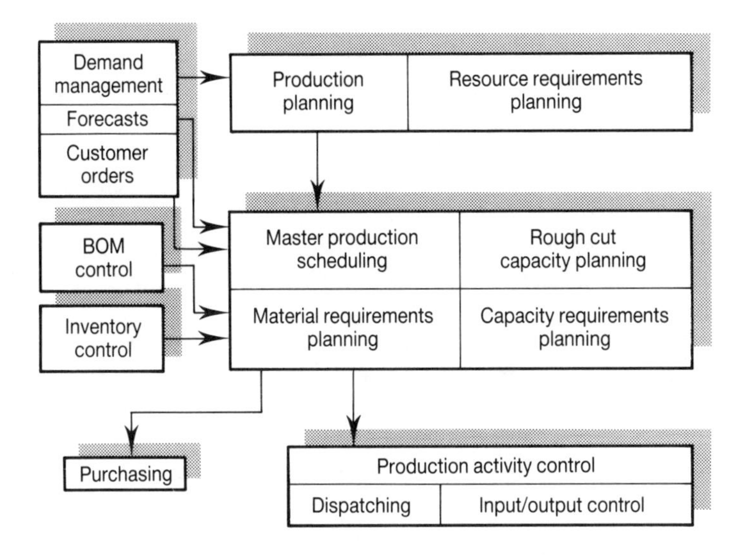

Figure 7.2 Manufacturing resource planning.

The modular structure of a typical MRP II system is shown in Figure 7.2.

In Chapter 4 it was seen that the production management system, as exemplified by an MRP II system, can be considered a significant island of automation in manufacturing. This is attested to by its integration of many diverse manufacturing functions, from financial systems down to the shop floor. However, as an island of automation, it has several inadequacies in terms of broader integration into CIM.

A brief description of the various modules is given in the following sections.

7.3 Master production scheduling

The MPS is a statement of what the company plans to manufacture. It is the planned build schedule, by quantity and date, for top level items, either finished products or high level configurations of material (either physical configurations or pseudoconfigurations used solely for planning purposes). The MPS module takes into account the sales forecast as well as considerations, such as backlog, availability of material, availability of capacity, management policy and company goals, etc. in determining the best manufacturing strategy.

The MPS *drives* MRP and is thus the key input into the MRP process. Any errors within it, such as poor forecasts, etc. cannot be compensated for by sophisticated MRP analysis as in lot sizing, calculation of safety stock or rescheduling. The MPS must be realistic in terms of the goals it sets for the manufacturing facility. It must not be merely a *wish list* of desirable production levels set by top management.

The accuracy of the MPS varies over the planning horizon. The planning data for the near term would tend to be more accurate since it is dominated by actual customer orders, distribution warehouse requirements and spare parts requirements. Change to the short term MPS should be a rare occurrence and the short term MPS should be treated as a series of firm planned orders.

Further out in the planning horizon, the MPS is likely to be less accurate and to be dominated by forecasts rather than actual orders. Forecasts may be based on analysis of historical trends, consideration of the state of the economy and market, and the actions of competitors. It may reflect the *best guesses* of those close to the market or it may involve the use of analytic forecasting and trend analysis techniques. Techniques, such as moving average analysis, exponential smoothing and regression analysis, may well be used to analyze past data in order to predict future data. The person or group responsible for the forecast must be aware of where each product in the company's portfolio of products sits in terms of the product life cycle and take this into account when preparing a forecast. This latter is particularly important today in industries, such as electronics and telecommunications, where product life cycles tend to be relatively short.

A typical MPS might look like Table 7.1. Typically the master schedule module software allows for a system generated forecast, a manually entered forecast, a schedule of actual customer orders received and a very simple procedure to combine the above into a working estimate of demand. The user may specify a master production schedule. Each order is treated as a firm planned manufacturing order. There is a netting process very similar to MRP as the forecast demand is netted with the MPS and current inventory to generate a projection of inventory on-hand and **available to promise**. Projected inventory is based on initial inventory plus the firm planned orders less total demand. Available to promise is based on initial inventory plus firm planned orders less actual orders.

The difference between the MPS module and the MRP procedure is that demand will only propagate from the scheduled MPS and not from the projected requirements. This means that the MPS will not influence manufacturing or purchasing orders without the intervention of the master planner.

In Table 7.2 the full MPS planning analysis for Stool B is illustrated. The aggregating procedure for demand in this instance is as follows: in the first 2 weeks the actual orders dominate while beyond this time, manual forecasts dominate. In the case where there is no manual forecast the system generated forecast dominates.

Table 7.1 The first 13 weeks of the MPS.

	Month 1				**Month 2**				**Month 3**				
	1	*2*	*3*	*4*	*5*	*6*	*7*	*8*	*9*	*10*	*11*	*12*	*13*
Stool A	70	70	70	70	70	70	70	70	70	70	70	70	70
Stool B	80	80	80	80	80	60	60	60	60	60	0	0	0
Stool C	100	100	120	120	120	120	140	140	140	140	160	160	160
Stool D	10	15	15	10	10	10	15	15	10	10	10	10	10
Stool E	40	40	35	35	30	30	30	20	20	20	20	20	20

Table 7.2 MPS analysis of Stool B.

	Item: Stool B				**Part number: F-449**								
Week number	*1*	*2*	*3*	*4*	*5*	*6*	*7*	*8*	*9*	*10*	*11*	*12*	*13*
System forecast	60	60	60	60	60	60	60	60	60	60	60	60	60
Manual forecast	50	60	40	50	70	80	80	70	70				
Actual orders	55	60	30	20	25	5		10					
Total demand	55	60	40	50	70	80	80	70	70	60	60	60	60
Firm planned orders	80	80	80	80	80	60	60	60	60	60			
Net requirements													55
Starting inventory	60												
Projected inventory	85	105	145	175	185	165	145	135	125	125	65	5	−55
Cumulative available to promise	85	105	155	215	270	325	385	435	495	555	555	555	555

It is important that there be a check on the feasibility of the proposed MPS before it is frozen and released to the manufacturing system for implementation. This feasibility check may be carried out through rough cut capacity planning. Master planning and rough cut capacity planning are thus two techniques that are employed in parallel. Actual MRP II systems vary somewhat in the support they give to these functions. Master planning systems often allow for planning at multiple levels and similar techniques to MPS and RCCP can be applied at the more aggregate business planning and production planning levels. The whole area of master planning is well treated by Berry *et al.* (1979) and also by Vollman *et al.* (1984).

7.4 Rough cut capacity planning

Rough cut capacity planning involves a relatively quick check on a few key resources required to implement the master schedule, in order to ensure that the MPS is feasible from a capacity point of view. The MPS and the rough cut capacity requirements plan are developed interactively.

In rough cut capacity planning, a bill of resource is attached to each of the master scheduled items. This bill of resource describes the capacity of various key facilities and/or people required to produce one unit of the item. Provision is made for using lead time offsets. No consideration is given to component inventories and the capacity requirements plan is driven solely by exploding the MPS against the bill of resource. The technique thus determines the impact of the master production schedule or the production plan on key or aggregate resources, such as man hours, machine hours, storage, standard cost dollars, shipping dollars, inventory levels, etc.

If the rough cut capacity planning exercise reveals that the MPS, as proposed, is infeasible then either the master production schedule must be revised or, alternatively, more resources must be acquired. Long term planning of overtime or subcontract is thus possible with the procedure.

Here again, the concept will be illustrated by an example. We are checking total man hours required to produce the master production schedule. Returning to the MPS example of Table 7.1, let us assume each stool requires 1 man hour of resource. A bill of resource for Stool A through E is illustrated in Table 7.3. We are ignoring any lead time offset. Each week there are a total of 300 man hours available in the factory. The rough cut capacity plan is illustrated in Table 7.4. The source of capacity requirements is illustrated in Table 7.5.

7.5 Capacity requirements planning

The MRP system produces a set of planned orders for both manufactured and purchased items to meet the requirements of the master schedule. The MRP system has generated this material plan based on planned lead times for

Table 7.3 Bill of resource for Stools A through E.

STOOL A

Part number	Part description	Resource description	Resource quantity
F-456	Stool A	Man hours	1.0

STOOL B

Part number	Part description	Resource description	Resource quantity
F-449	Stool B	Man hours	1.0

STOOL C

Part number	Part description	Resource description	Resource quantity
F-431	Stool C	Man hours	1.0

STOOL D

Part number	Part description	Resource description	Resource quantity
F-426	Stool D	Man hours	1.0

STOOL E

Part number	Part description	Resource description	Resource quantity
F-412	Stool B	Man hours	1.0

manufactured and purchased items. In so doing, the system has ignored any capacity constraints in the manufacturing facility.

MRP output can, however, be used for capacity requirements planning. This is done by exploding the manufacturing orders (planned and actual) through the routing specified in the production activity control system. This generates a detailed profile of what capacity is required in each work centre.

Table 7.4 Rough cut analysis.

Resource: Total man hours													
Week	*1*	*2*	*3*	*4*	*5*	*6*	*7*	*8*	*9*	*10*	*11*	*12*	*13*
Required std. hours	300	305	320	315	310	290	315	305	300	300	260	260	260
Available std. hours	300	300	300	300	300	300	300	300	300	300	300	300	300
− Deviation		−5	−20	−15	−10		−15	−5					
+ Deviation						+10					+40	+40	+40

Table 7.5 Source of resource requirements.

Item: Total man hours

Week	*1*	*2*	*3*	*4*	*5*	*6*	*7*	*8*	*9*	*10*	*11*	*12*	*13*
Stool A	70	70	70	70	70	70	70	70	70	70	70	70	70
Stool B	80	80	80	80	80	60	60	60	60	60	0	0	0
Stool C	100	100	120	120	120	120	140	140	140	140	160	160	160
Stool D	10	15	15	10	10	10	15	15	10	10	10	10	10
Stool E	40	40	35	35	30	30	30	20	20	20	20	20	20

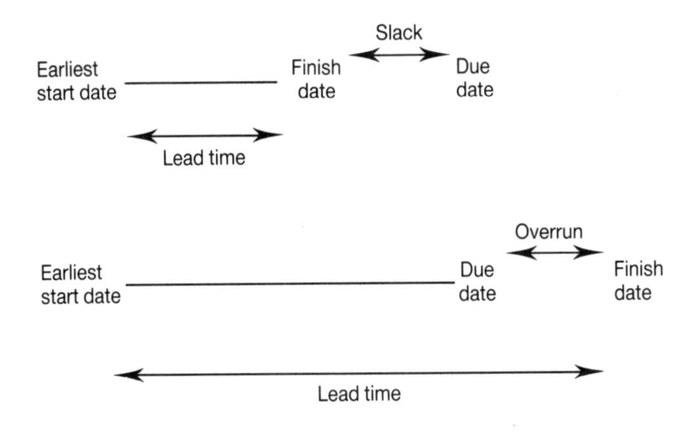

Figure 7.3 Forward scheduling.

Required capacity is then compared with available capacity and overload/underload conditions are identified.

Capacity requirements planning generates a more detailed capacity profile than that generated by RCCP. However, CRP is only performed after each MRP run and these runs are typically performed only once a week. Therefore CRP does not facilitate interactive planning and is used primarily as a verification tool. In the days when the notion of closed loop MRP was popular, the CRP module was emphasized but, since then, its role seems to have diminished in importance when compared to RCCP.

There are two common capacity planning methods, namely capacity planning based on **forward scheduling** and capacity planning based on **backward scheduling**. The principle of forward scheduling is illustrated in Figure 7.3. We identify the earliest start time for the operation in question through consideration of the planned order release date generated by MRP. We then use the lead time to calculate the completion date. If the completion time is prior to the due date then we have slack time available. If the completion time is after the due date then we have a delay.

In backward scheduling we identify the due date for the operation in question and use the operation lead time to calculate the latest operation start date. If the latest start time is after the earliest start date then we have slack time available. If the operation start time is prior to the earliest start date then we have negative slack. This is illustrated in Figure 7.4.

From the production activity control system we have available the planned operation time for each operation in the manufacturing routing, as well as the details of the work centre at which the work will be carried out. What CRP does is to develop a load profile for each work centre which describes what capacity is required over the planning horizon and what

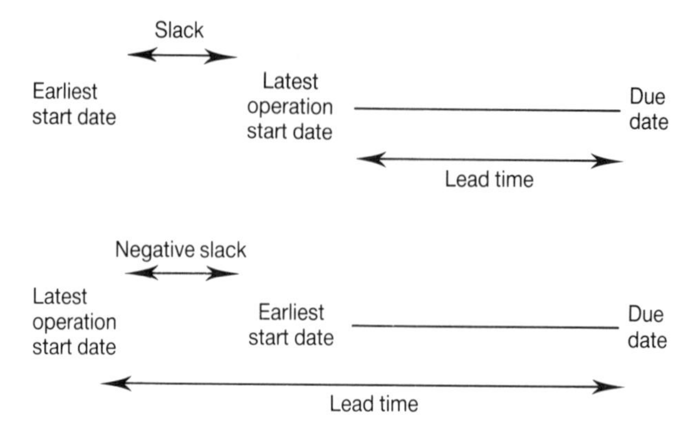

Figure 7.4 Backward scheduling.

capacity is available. Some CRP systems ignore operation lead times and use only the planning lead time for the order from MRP. In any case, the principle is the same. In our CRP example we are using operation detail from the PAC system to drive CRP.

The **operation lead time** referred to here is to be distinguished from the **planning lead time** used by MRP. The planned lead time, as used by MRP, is the average lead time for the completed component. In manufacturing a component or sub-assembly, several operations may be required, for example assembly, painting and inspection, to complete the item. Each of these operations has associated with it, in the PAC system, an operation lead time. These operation lead times when added together make up the planning lead time for the component, as used by MRP.

We will use the example of the stool, which we discussed earlier, to illustrate this. Consider again the assembly of Stool A, assuming that the requisite components – the legs and the seats – are available. Table 7.6 presents the routing information for assembling Stool A while Table 7.7 presents the work centre information.

If we assume a batch size of 20 then the calculation of the operation lead times is as illustrated in Table 7.8.

The item lead time for Stool A is the sum of these three operation times, namely 76.3 hours or approximately two weeks, assuming a 40 hour working week. Hence the figure of two weeks lead time attributed to Stool A and stored with the master parts information.

We note that the estimated lead time is a function of the batch size. If we doubled the batch size, i.e. increased it to 40, there would not be a significant increase in the estimated lead time, since only the process time component of lead time increases. In fact, a simple calculation will show that doubling the batch size increases the component lead time to 92.3. Thus, although the lead time is a function of the batch size, the fact that, in conventional manufacturing systems, queue time typically occupies the

Table 7.6 Routing information for stool A.

Operation number	10
Description	Assemble legs to seat
Set-up time	0.5 hours
Processing time	0.25 hours
Operator time	0.25 hours
Transport time	1.00 hours
Work centre	Assembly shop
Next operation	20
Operation number	20
Description	Paint the stool
Set-up time	0.75 hours
Processing time	0.35 hours
Operator time	0.35 hours
Transport time	1.00 hours
Work centre	Paint shop
Next operation	30
Operation number	30
Description	Inspect the stool
Set-up time	0.05 hours
Processing time	0.20 hours
Operator time	0.20 hours
Transport time	1.00 hours
Work centre	Inspection laboratory
Next operation	Stock room

lion's share of manufacturing lead times allows MRP analysts to assume that the lead time is, for all practical purposes, independent of the batch size in use.

Given the importance of lead times within MRP let us now review the lead time issue in some detail. The APICS dictionary (Wallace 1980) defines lead time as follows:

> 'A span of time required to perform an activity. In a production and inventory context, the activity in question is normally the procurement of materials and/or products either from an outside supplier or from one's own manufacturing facility. The individual components of any given lead time can include some or all of the following: order preparation time, queue time, move or transportation time, receiving and inspection time'.

The manufacturing lead time is 'the total time required to manufacture an item. Included here are order preparation time, queue time, set-up time, run time, move time, inspection and put away time'.

Table 7.7 Work centre information.

Work centre	Assembly shop
Queue time	8.0 hours
Available capacity	40.0 hours
Work centre	Paint shop
Queue time	16.0 hours
Available capacity	40.0 hours
Work centre	Inspection laboratory
Queue time	32.0 hours
Available capacity	40.0 hours

Table 7.8 Calculation of operation leadtimes.

Operation number: 10 Assembly	
Queue time	8.00 hours
Set-up time	0.50 hours
Processing time × Batch size	
(0.25 hours × 20)	5.00 hours
Transport time	1.00 hours
Total operation time	14.50 hours
Operation number: 20 Painting	
Queue time	16.00 hours
Set-up time	0.75 hours
Processing time × Batch size	
(0.35 hours × 20)	7.00 hours
Transport time	1.00 hours
Total operation time	24.75 hours
Operation number: 30 Inspection	
Queue time	32.00 hours
Set-up time	0.05 hours
Processing time × Batch size	
(0.20 hours × 20)	4.00 hours
Transport time	1.00 hours
Total operation time	37.05 hours

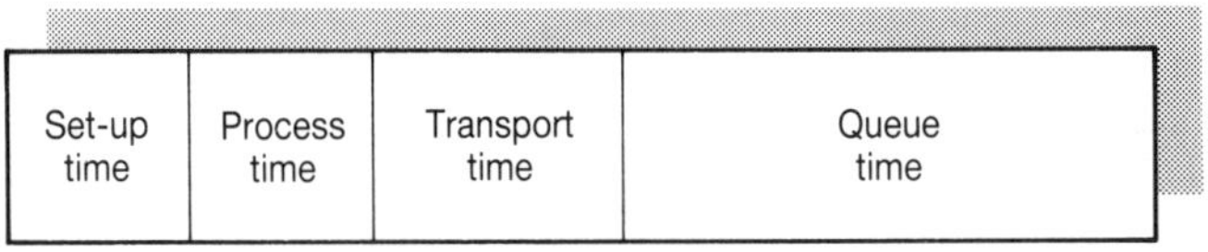

Set-up time	Process time	Transport time	Queue time

Figure 7.5 Breakdown of the lead time in a batch production system.

The purchasing lead time is 'the total time required to obtain a purchased item. Included here are procurement lead time, vendor lead time, transportation time, receiving, inspection and put away time'.

Lead times must then be determined for both purchased and manufactured items. Lead times for purchased items are determined following discussions and negotiation between the purchasing people within a company and suppliers. Lead times for manufactured items, on the other hand, must be estimated based on past experience and through consideration of the various elements which together make up total lead time.

Within batch production systems, the lead time or throughput time for a batch through the shop floor is typically much greater than the processing time for the batch as shown in Figure 7.5. It is not unusual for the actual processing (including set-up time) to represent less than 5% of the total throughput time. The throughput time or lead time is made up mainly of four major components – the set-up time, the process time including inspection process time, the transport time and the queuing time. In real life this latter component is the largest, often representing in excess of 80% of total throughput time.

In estimating lead times we must therefore pay particular attention to the calculation of the queuing element of the lead time. It is clear that this queuing time depends greatly on the load in front of a work centre at a given point in time. Thus, if the machine is idle with no work queued in front of it, the queue time for an arriving job will be almost zero. If the machine is busy with a large queue of work then, depending on the priority of the job in question, the queue time could be quite large. It is clear that the actual queuing time is variable and, consequently, the lead time is a variable also. For capacity planning purposes we use an average lead time based on our experience of the flow of work through the shop floor. Hoyt (1983) suggests that the lead times should be determined dynamically and offers a simple method for their determination. The suggestion was that the lead time for a given work centre is 'the most recent production period's average queue divided by its average output'. Thus, if the average queue of work in front of a work centre is 600 hours

Table 7.9 Planned orders for Stool A.

Release date	Due date	Planned orders	Number of batches of 20
Week 5	Week 7	40	2
Week 8	Week 10	80	4

Table 7.10 Calculation of department loads.

Assembly department

Load from Stool A
(Batch size × Processing time) + Set-up
40 × 0.25 + 0.5 = 10.5 hours

Painting department

Load from Stool A
(Batch size × Processing time) + Set-up
40 × 0.35 + 0.75 = 14.75 hours

Inspection department

Load from Stool A
(Batch size × Processing time) + Set-up
40 × 0.20 + 0.05 = 8.05 hours

and the work centre's average output per week is 200 hours then the lead time, based on this method, is three working weeks or 15 days. As we shall see in Chapter 10 and Part IV of this book, this attitude to lead times is seen by many people as a fundamental weakness in the MRP approach.

Returning to our example, the planned orders for Stool A for our example MRP output are as in Table 7.9.

The load on the three departments arising from the planned order release of 40 of Stool A at the end of week 5 can now be calculated as in Table 7.10.

Based on the planned orders defined above, the load projection for each department is given in Figures 7.6 through 7.8.

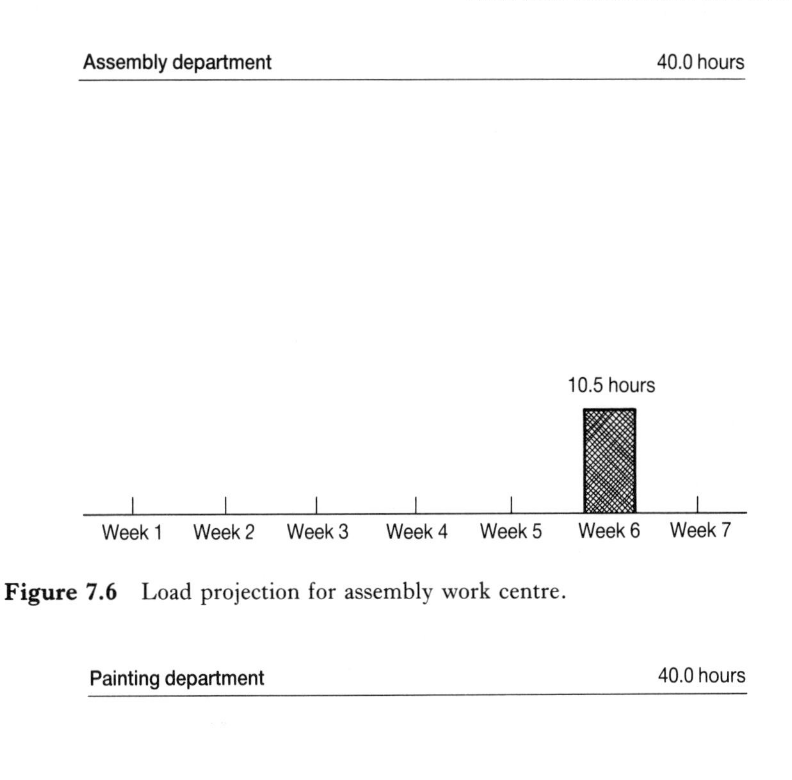

Figure 7.6 Load projection for assembly work centre.

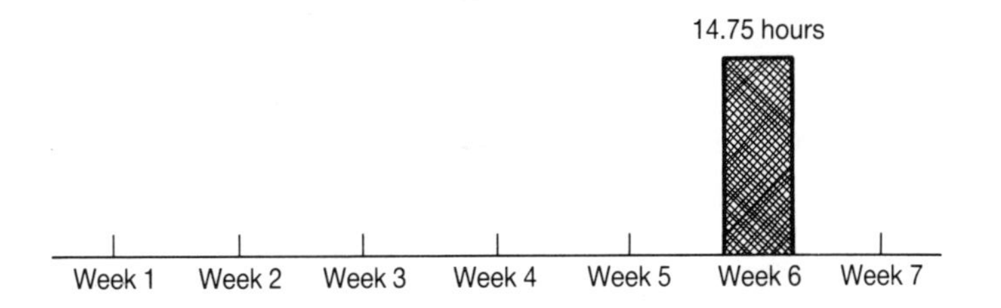

Figure 7.7 Load projection for the painting work centre.

Note that both the assembly and painting operations have been scheduled for week 6 due to lead time considerations, while the inspection was loaded in week 7. This is the effect of the operation scheduling procedure.

Assume we are using backward scheduling. (If we use forward scheduling we get more or less the same result.) The backward scheduling procedure works as follows. The finish of the last operation (inspection) is scheduled for the due date, i.e. end of week 7. The operation lead time for the inspection

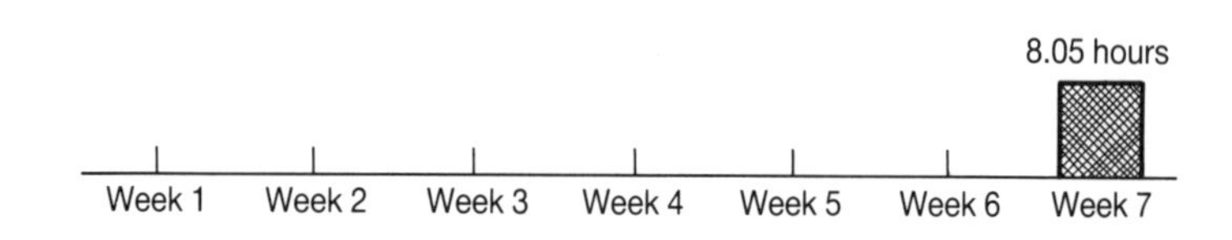

Figure 7.8 Load projection for the inspection work centre.

operation is 41.05 hours for a batch of 40. Therefore, the previous operation (painting) is scheduled for the previous week (week 6). The operation lead time for the painting operation is 31.75 hours for a batch of 40, which means that the assembly operation can be scheduled for completion in week 6 also.

The above example illustrates the build-up of load projections using the set-up and processing times to determine the machine or department hours and, secondly, the operation times to understand the lead times between operations.

In a similar manner to the above, we can calculate the capacity requirements arising from all planned orders for all products and thus build up a comprehensive picture of the load on the production departments arising from the MRP planned output. In this way, the feasibility of the schedule from the point of view of capacity can be studied, bottlenecks can be identified and overtime or subcontract planned, if necessary.

A typical load projection arising from CRP appears in Figure 7.9. This shows a projected overload in the assembly department in week 2 and a low load in week 7.

7.6 Production activity control

At the production activity control level, functions have been added to the basic MRP system which describe the process routing for fulfilling a manufacturing order (i.e. the sequence of steps a part goes through in its

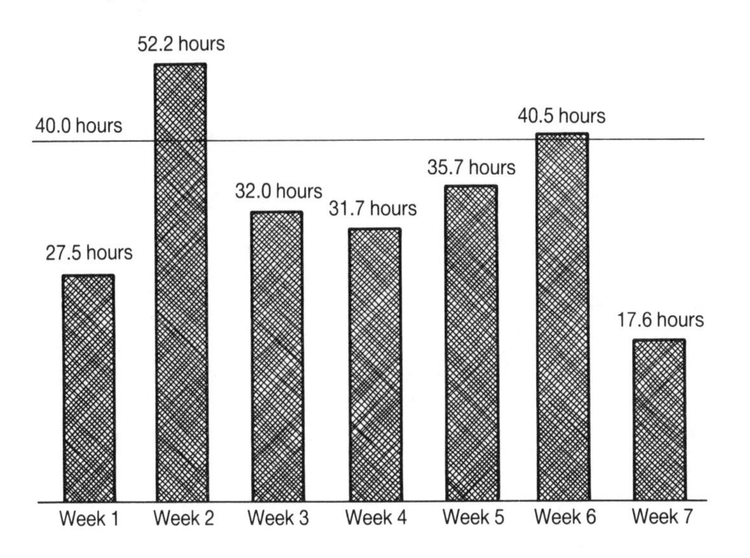

Figure 7.9 Load projection for the assembly work centre.

manufacturing process), as well as work centre and standard time information. The routing information and work centre information of the PAC system have already been described in Section 7.5.

The same operation scheduling procedures, which we saw in the CRP system, are used to set planned operation start times on dispatch lists. Work in progress tracking facilitates the prioritizing of manufacturing orders on the dispatch list. There are a variety of techniques, including the use of critical ratio techniques, available for generating this priority. The critical ratio is calculated as the ratio of the the time remaining until the due date, as compared with the sum of operation times remaining.

Capacity control, in the form of input output control, is provided in some systems. This reports on the planned workload that should flow into and out of each work centre by week, as well as the actual workload inflow and outflow and, finally, the cumulative deviation between plan and actual. This is useful in determining if a problem in a work centre's output is caused by a capacity problem at that work centre or, alternatively, by late input from upstream work centres.

Some of the inadequacies of PAC modules in MRP II systems were described in Chapter 4. Some of these weaknesses are again listed:

- Absence of *off the shelf* interfaces to automatic data collection devices.
- Absence of quality management functionality.

- The routing control functionality of PAC modules in MRP II systems is typically rather naïve. Only straight line flows are typically allowed and there is no support of multilevel tracking, i.e. tracking both the top level product and key sub-assemblies and components through parallel routings. In addition, tracking is primarily by manufacturing order and support is rarely given for individual identification of units within a batch.
- Tooling control is rarely adequately provided for.
- Preventative maintenance and equipment tracking is rarely available.

7.7 Conclusion

This chapter has focused on MRP II systems, highlighting closed loop MRP, as well as several of the important modules that make up an MRP II system. Some of the data needed to operate an MRP II system will be discussed in Chapter 8.

CHAPTER EIGHT

The production database

8.1 The production database

It is clear from our discussion thus far that an MRP II, or indeed an MRP, system relies on a great deal of data. The intention in this section is to look in more detail at this data. We call it the production database. The intention is not to give an exhaustive list but merely to discuss, in general terms, the more important sources of data used and maintained by an MRP II system. The implementation of the database will not be discussed. MRP II systems are gradually migrating from file oriented data storage to database management systems. Suffice it to say that data should be stored in a manner that avoids redundant storage, with links between related fields being system maintained, and which facilitates the ease of access in any desired manner (either for inquiry or reporting purposes) or by applications external to the MRP II system.

A typical production database contains several related major sources of information including:

- The master parts information.
- Full inventory information.
- Bill of materials information.
- The manufacturing routing.
- Work centre information.
- Tooling information.

We have already touched upon nearly all of these information sources. We will now review each source in turn and indicate the type of data it contains.

Table 8.1 Master parts information on each item.

• Part number	– a unique identifier for each item
• Part description	
• Unit of measurement	
• Lot size policy	
• Lot size quantity	
• Safety stock	– if relevant
• Shrinkage factor	– if relevant
• Lead time	
• Safety lead time	
• Make or buy code	
• Supplier code	
• Stores location	– if relevant
• MPS code	– indicates whether this is a master planned part
• Low level code	– indicates the lowest level on any BOM that a part is to be found
• Standard cost	– if relevant
• Material cost	– if relevant
• Labour overhead	– if relevant
• Overhead cost	– if relevant

8.2 The master parts information

Master parts information contains detailed data on each planned item in the MRP II system. Each part is typically described in the terms given in Table 8.1.

The data in Table 8.1 is what might loosely be termed **static data**, in the sense that the individual data values do not change very frequently. Furthermore, changes to any of these data values are typically initiated by the user, as distinct from being derived from calculations within the MRP explosion procedure. The user has software support in lot size determination and in low level code maintenance.

The majority of the data fields listed are self explanatory. Lot sizing procedures will be discussed in Chapter 9. The lead times are predetermined and used to offset the purchasing and manufacturing orders, and hence generate a schedule. In many systems the possibility exists to capture data directly from the shop floor or from the purchasing system on actual lead times. This facility supports the user in maintaining up-to-date lead times. The shrinkage factor allows the system to account for likely stock wastage, scrap or obsolescence.

Table 8.2 Full inventory information.

- Current inventory (by location)
- Allocations
- Open orders per period/time bucket over the planning horizon

- Gross requirements per period/time bucket over the planning horizon
- Net requirements per period/time bucket over the planning horizon
- Planned order releases per period/time bucket over the planning horizon

Table 8.3 Bill of materials information.

- Parent part number

- For each component:
  - the component part number
  - the quantity required

8.3 Full inventory status

The second data source (Table 8.2) can be termed **dynamic**, in that the values stored within the data fields change very frequently and many are generated as a consequence of MRP calculations. This data describes the full inventory status for the part, including requirements, allocations, open orders and planned orders. As noted in Chapter 6, it is necessary for a net change system to store full requirements information. In a minimal regenerative MRP system, it is possible merely to retain inventory data in the narrow sense of stock, allocations and open orders data.

The full inventory information may also include data to support master scheduling, as described in Tables 7.1 and 7.2 of Chapter 7.

8.4 Bill of materials information

The bill of materials (see Table 8.3) defines the structure of a product. Data is stored in a manner that supports the BOM inquiry options normally available in a bill of materials system. Such options include the ability to generate single level assembly BOMs, indented bills of materials, summary bills of materials and *where used* tables for individual assemblies and components.

Table 8.4 Routing information.

- Part number
- For each operation:
  - the operation number
  - a description of the operation in question
  - the work centre at which the operation should be performed
  - an alternative operation to do the same task
  - a reference to the manufacturing documentation associated with the operation
  - the tools required for operation
  - the set-up time of the job on the machine/work centre
  - the processing time of the job on the machine/work centre
  - the operator time for the job in question
  - the time to transport the batch or job to the work centre
  - the operation lead time

8.5 Routing information

Routing information defines the manufacturing and/or assembly operations which must be performed on a manufactured component. Its use has already been discussed in Chapter 7. The data is primarily of an engineering nature. Typical data maintained on a manufactured parts routing is shown in Table 8.4.

8.6 Work centre information

Work centre information is used primarily for capacity planning purposes. It contains data on each work centre in the production facility. In this context a *work centre* is a set of resources. Thus, it may refer to a group of machines and/or operators with identical functionality and ability to discharge that functionality or it may refer to a single resource. A typical work centre description is outlined in Table 8.5.

8.7 Tooling information

Tooling information provides detailed data on tools which are available and are associated with particular operations and work centres. For a company engaged in substantial metal cutting or metal forming operations, one would certainly expect to find great emphasis on such information. Less emphasis would be placed on the same information by a firm engaged primarily in, say, electronics assembly. Table 8.6 shows what a typical description might include.

Table 8.5 Work centre information.

- Work centre number
- Work centre description

- Available capacity
- Units of capacity

- Normal queue time at the work centre

- Work centre costs
  - labour cost per unit time
  - machine cost per unit time
  - overhead cost per unit time

Table 8.6 Tool information.

- Tool number
- Tool name
- Tool description
- Tool drawing number if relevant
- Tool location in stores
- Tool status
- Alternative tool if available
- Tool life
- Accumulated tool life worked
- Unit for tool life calculations

8.8 Conclusion

An attempt has been made, in very general terms, to describe some of the information required to support the operation of an MRP II system. The implementation of this production database was not discussed. Lot sizing techniques in MRP systems will be discussed in Chapter 9. Readers may, if they so desire, skip directly to Chapter 10 without loss of understanding.

CHAPTER NINE

Lot sizing in MRP systems

9.1 Introduction

Most of the examples on MRP that have been presented in this book have simply taken the offset net requirements to constitute the planned order schedule. However, there are many situations where constraints on the order lot size make this an unsuitable procedure. For purchased items, vendors may supply only in multiples of a given number and the net requirements may have to be batched so as to accommodate this. Similarly for manufacturing, a process involving high set-up costs may dictate the use of a definite lot size policy.

An MRP system must, therefore, accommodate a procedure which facilitates the calculation of lot sizes on some basis other than simple acceptance of the values that fall out from the net requirements calculation. Furthermore, this procedure must be embedded within the MRP explosion procedure since the order size for a parent item determines the gross requirements for its components. Lot sizing decisions made high up the BOM structure will have ripple effects right down through the planning of all components in the bill of materials.

A number of procedures are available to help determine the appropriate lot size. These range from relatively simple procedures to very complicated algorithms. A representative selection of the procedures will be reviewed here including:

- The lot for lot method.
- The fixed order quantity method.
- The economic order quantity method.

Table 9.1 Analysis of Stool A using lot for lot technique.

Item: Stool A	**Part number: F-456**									
Week number	*1*	*2*	*3*	*4*	*5*	*6*	*7*	*8*	*9*	*10*
Gross requirements							50			80
Scheduled receipts										
Projected inventory	10						−40			−120
Net requirements							40			80
Planned orders					40			80		

- The method of fixed order periods.
- The periodic order quantity method.
- The method of part period balancing.
- The Wagner Whitin algorithm.

The Wagner Whitin algorithm is discussed since it results in an *optimum* ordering policy, provided the relevant assumptions hold. The other procedures listed here are more in the form of heuristics which give good or poor results depending on the set of data in question. Other lot sizing methods, such as the Gaither method (1981), the modified Gaither method (1983), the Silver–Meal technique (1973) or the method proposed by Groff (1979) will not be covered. The interested reader may pursue the references at the end of Part II.

9.2 The lot for lot method

Lot for lot is the simplest of the lot sizing methods and involves the direct translation of net requirements into order quantities. It is, in effect, the method used in almost all of the MRP examples thus far. For each net requirement in each period there is an order offset by the appropriate lead time as in Table 9.1.

Table 9.2 Analysis of Stool A using fixed lot size.

Item: Stool A	**Part number: F-456**									
Week number	*1*	*2*	*3*	*4*	*5*	*6*	*7*	*8*	*9*	*10*
Gross requirements							50			80
Scheduled receipts										
Projected inventory	10						−40			−120
Net requirements							40			80
Planned order receipt							200			
Planned orders					200					

9.3 The fixed order quantity method

This method is quite frequently used in practice. The net requirements are checked against the assigned fixed lot size. If the net requirements were less than or equal to the lot size, then the amount specified in the lot size is ordered. Otherwise the order size is equal to the net requirements. Taking the example of Stool A and assuming that the fixed order quantity for this item is 200, we arrive at the situation depicted in Table 9.2.

Note that another line has been added to the planning sheet, which indicates the receipt of planned orders. This is done in order to see the effect of lot sizing independently of the application of the lead time offset.

The fixed lot size quantity may be set for an item based on local constraints around packaging, material handling or, alternatively, may be calculated by the economic order quantity analysis described in Section 9.4.

9.4 The economic order quantity method

Large batch sizes result in high inventory levels which are, of course, expensive in terms of the cost of capital tied up in inventory. Small batches imply a proportionately lower inventory cost. However, there is a set-up cost incurred with the placing of an order or the start-up of a batch on a machine.

Table 9.3 Calculation of EOQ.

Let S = set-up cost per batch
Let C = inventory carrying cost per item per unit time
Let Q = the batch size
Let D = the demand for the item per unit time
Let TC = the total cost of inventory and set-up

Average inventory level = $Q/2$
Inventory cost per unit time = $CQ/2$
Set-up cost per unit time = SD/Q

Therefore TC = $(CQ/2 + SD/Q)$

To minimize the total cost (TC) with respect to Q we simply differentiate with respect to Q:

$$\frac{\mathrm{d}TC}{\mathrm{d}Q} = (C/2 - SD/Q{**}2)$$

Equating this first derivative to zero gives us a point of inflection, in this case the minimum cost batch size:

$$(C/2 - SD/Q{**}2) = 0$$

$$=> \quad Q{**}2 = 2SD/C$$

$$=> \quad Q = \sqrt{2SD/C}$$

This set-up cost (for manufactured items) or ordering cost (for purchased items) must be amortized over the batch or order size. If set-up or ordering costs are high then we may need to resort to larger batches to reduce the per unit cost of set-up and thereby incur larger inventory costs. It is clear, therefore, that there is a trade off between order (or set-up costs) and inventory costs. The Economic Order Quantity (EOQ) formula is simply a mathematical expression of this tradeoff and reflects the minimum total cost of carrying stock and set-up. The EOQ is also known as the EBQ (Economic Batch Quantity).

The derivation of the simple EOQ model is illustrated in Table 9.3. There are some important assumptions underlying this calculation, namely:

(1) Demand for the item in question is known and constant.

(2) The set-up cost and the inventory carrying cost are known.

Thus we have a value of Q, the order size that is optimum and reflects the most effective tradeoff between set-up and inventory costs.

This type of analysis can be used for lot sizing in MRP systems, provided we understand the basic assumptions inherent in the calculation and use the result with care. For example, the assumption of constant demand for an item is clearly not true in most MRP installations. Demand in

Table 9.4 Numerical example of EOQ calculation.

$$EOQ = \sqrt{\frac{2 \times 150 \times 676}{10}}$$

$$=> EOQ = 142$$

Table 9.5 Analysis of Stool A using EOQ.

Item: Stool A	**Part number: F-456**									
Week number	*1*	*2*	*3*	*4*	*5*	*6*	*7*	*8*	*9*	*10*
Gross requirements							50			80
Scheduled receipts										
Projected inventory	10						−40			−120
Net requirements							40			80
Planned order receipt							142			
Planned orders					142					

MRP type situations tends to be non-uniform, in fact, lumpy. Furthermore, it is often difficult to calculate accurately the cost of set-up and carrying inventory. However, the greatest difficulty in applying EOQ in manufacturing situations stems from the fact that the majority of parts in a BOM are dependent parts and cannot be considered in isolation. The EOQ formulation implicitly assumes that demand for an item is independent and that its batch size need not take into account the demand for or the batch sizes of other items.

As an example, consider Stool A once more. The demand for this item is 130 over a ten week period which represents an annual demand of 676. Let us assume that the set-up cost is \$150 and that the inventory carrying cost is \$10 per unit per year. The EOQ is calculated as in Table 9.4.

The revised planned order schedule is as in Table 9.5.

Table 9.6 Relationship between set up cost and EOQ.

Set-up cost = \$150	=> EOQ =	142
Set-up cost = \$15	=> EOQ =	45
Set-up cost = \$1.5	=> EOQ =	14

Perhaps the most important point about the EOQ formula is not whether it can be used directly within an MRP environment, but rather the insights it offers. In particular, this formula illustrates how the batch size increases with higher set-up cost.

Taking the example of the Stool A referred to above we can see how a reduction in the set-up cost reduces the economic batch quantity in Table 9.6.

Clearly, in a situation where flexibility is of key importance, we must seek to achieve small batches and ultimately a batch size of one. A *liability* of MRP is that, as an *island of automation*, it seeks to generate a schedule for purchased and manufactured parts and takes data, such as set-up time/cost, as *given*. It does not focus our attention on extremely important issues, such as the possibility of *reducing set-up cost*.

9.5 The fixed order period method

This method is somewhat similar to the fixed quantity approach in that it sets a fixed time between orders (as distinct from a fixed order size) and orders the amount required to meet the demand in that period. In effect, the ordering policy is saying *order X weeks supply* where X is determined for the part being planned. Weeks with no net requirement are passed over and not counted as part of the ordering interval.

Continuing our example using Stool A, if we set a fixed order period of 4 weeks our planned orders would be as in Table 9.7.

The order period can be set on an *ad hoc* basis or, perhaps, calculated on the similar basis to the EOQ described above. This latter option is termed the periodic order quantity (POQ).

9.6 Periodic order quantity

Periodic Order Quantity (POQ) is a variation of the fixed order period method discussed above, where the ideas from EOQ are used to calculate the time between orders. EOQ leads to a fixed order quantity and a variable interval between orders, while the periodic order quantity approach leads to variable order sizes with a fixed and constant time interval between orders.

Table 9.7 Analysis of Stool A using fixed order period.

Item: Stool A	**Part number: F-456**									
Week number	*1*	*2*	*3*	*4*	*5*	*6*	*7*	*8*	*9*	*10*
Gross requirements							50			80
Scheduled receipts										
Projected inventory	10						−40			−120
Net requirements							40			80
Planned order receipt							120			
Planned orders					120					

As discussed earlier, EOQ gives the *optimum* order quantity for a given independent item based on a known set-up cost, known cost of carrying inventory and a known demand level. In the case of the periodic order quantity, the time between orders is calculated by dividing the demand per period by the EOQ. In the case of our example of Stool A, the demand per year is 676 and the EOQ is 142, which suggests approximately 5 orders per year and a time interval between orders of 11 weeks.

In our ongoing example, POQ leads to the situation depicted in Table 9.8.

9.7 Part period balancing

This method also stems from the thinking behind the EOQ formula. The technique, as described by Berry (1972) and DeMatteis and Mendoza (1968), is probably best explained through an example. It seeks to equate the cost of set-up/order placement with the cost of inventory. It is based on the observation that the sum of the set-up/ordering costs and the inventory costs in the EOQ formula area minimized at the point at which the two costs are equal.

Table 9.8 Analysis of Stool A using POQ.

Item: Stool A	**Part number: F-456**									
Week number	*1*	*2*	*3*	*4*	*5*	*6*	*7*	*8*	*9*	*10*
Gross requirements							50			80
Scheduled receipts										
Projected inventory	10						−40			−120
Net requirements							40			80
Planned order receipt							120			
Planned orders					120					

Table 9.9 Weekly requirements.

Week number	*1*	*2*	*3*	*4*	*5*	*6*	*7*	*8*	*9*	*10*
Net requirement	30	50	20	40	50	20	30	40	60	20

This *equality condition* is applied as meaning *more or less equal*. We say more or less equal for two reasons. Firstly we will have to use estimates for our values of set-up and inventory costs and, secondly, it is unlikely that we will have a ordering option available from our requirements profile which will give set-up costs equal to inventory costs.

The example which follows (Tables 9.9, 9.10 and 9.11) is based on the particular version of this procedure used by Berry (1972).

Let us assume set-up costs of $200 and inventory costs of $1 per unit per week. For the sake of the calculation we need to assume that the inventory is used up at a uniform rate over the period of the week. Thus, if 30 items are held in stock at the beginning of a week the inventory carrying cost will be (30 × ½ × 1) or $15.

Table 9.10 Part period balancing example.

Option 1 Order for week 1 only
Option 2 Order for weeks 1 and 2 only
Option 3 Order for weeks 1, 2 and 3 only
Option 4 Order for weeks 1, 2, 3 and 4 only

Option 1
Order policy : Order 30 units
Set-up cost = \$200
Inventory costs = $(30 \times \frac{1}{2} \times 1) = \$15$

Option 2
Order policy : Order 80 units
Set-up cost = \$200
Inventory costs = $((30 \times \frac{1}{2} \times 1) + (50 \times \frac{3}{2}))$
= \$90

Option 3
Order policy : Order 100 units
Set-up cost = \$200
Inventory costs = $((30 \times \frac{1}{2} \times 1) + (50 \times \frac{3}{2}) + (20 \times \frac{5}{2}))$
= \$140

Option 4
Order policy : Order 140 units
Set-up cost = \$200
Inventory costs = $((30 \times \frac{1}{2} \times 1) + (50 \times \frac{3}{2}) + (20 \times \frac{5}{2}) + (40 \times \frac{7}{2}))$
= \$280

Table 9.11 Planned receipts using part period balancing.

Week number	*1*	*2*	*3*	*4*	*5*	*6*	*7*	*8*	*9*	*10*
Net requirement	30	50	20	40	50	20	30	40	60	20
Planned receipts	100	–	–	140	–	–	–	120	–	–

Taking the position at the start of week 1, the various options and resulting analysis are illustrated in Table 9.10.

With option 4 the inventory costs exceed the set-up costs. Thus, somewhere between option 3 and option 4 there exists the crossover point where set-ups and inventory costs are equal. It is clear that option 3 is where the set-up costs most closely approximate to the inventory cost. Thus our decision is to avail of option 3. We then start a similar exercise with week 4 as our starting point. If we continue this exercise over the ten weeks of the planning horizon we will end up with the ordering strategy presented in Table 9.11.

9.8 The Wagner Whitin approach

This algorithm uses a dynamic programming approach to determine the optimum order quantities, given that the level of demand for a defined planning horizon is known and can be broken down into the discrete time periods and that the set-up and inventory carrying costs are known. The title of the original paper (Wagner and Whitin 1958) describes the approach as a dynamic version of the EOQ model. It is dynamic in the sense that it deals with demand that varies over a discrete horizon and generates variable lot sizes economically to satisfy that demand.

It is not normally used in practice because it is considered computationally too demanding from a data processing point of view and overly complex to be used in a manufacturing situation. With the increasing cost performance ratio of computer technology, the first argument is less true each passing year. The argument of complexity is valid although Fordyce and Webster (1984) have presented a useful tutorial paper which explains the thinking behind the method very well. Probably the greatest objection to Wagner Whitin is its claim to be optimal. While this is certainly true in a case of known demand, it is not so true when the set-up and inventory carrying costs are, at best, estimates and the statements of demand per time bucket in the planning horizon are less and less accurate the further out the planning horizon one looks.

9.9 How to choose the lot sizing policy

Berry (1972), in a study of lot sizing techniques, identified two sets of criteria for comparing lot sizing techniques. One set of criteria related to how easy the technique is to use in practice, how easy it is for production people to understand and how efficient it is in terms of computing time. The second set of criteria relate to its performance as a lot sizing technique in terms of the inventory and set-up costs it generates.

Unfortunately it is difficult to make a strong statement about the results of comparison studies they tend to be based on specific sets of data used in simulation models and are, therefore, only true in so far as the data used is representative of real life situations. For example, the study of De Bodt and Van Wassenhove (1983), which used one set of actual company data, reported that the EOQ rule performed well. Berry's work found that the performance of EOQ depended very much on the data set under analysis while the work of Silver and Meal (1973) indicates that EOQ is significantly worse than other techniques.

The lot sizing *problem* is the source of considerable divergence between those who practice MRP and many of those who are involved in research into production and inventory management systems. St. John (1984) puts in quiten succinctly when he says

> '. . . the marginal value of one more lot sizing comparison study is virtually worthless to the P&IC (Production and Inventory Control) practitioner, . . . plead with authors, researchers, software companies and practitioners alike to direct their attention to subjects that really need attention. There are some significant payoffs to be achieved in other areas that are going unexplored while some of the most brilliant minds in our profession continue to pollute the literature with more and more lot sizings studies.'

Mather (1985) offers a similar viewpoint and points out that a company is unlikely to be able to identify its set-up and work in progress costs accurately in order to plug them into the various lot sizing formulae. Mather also makes an interesting observation, which is particularly appropriate in the context of a CIM, namely:

> '. . . the question is, Why do we need to lot size? If we could remove the causes, then lot sizes could be small. Eliminating the causes completely would allow us to use the most economical lot size of all, one. The primary reasons for creating lot sizes are the cost of placing orders and the capacity lost during set-up. If these could both be reduced then lot sizes would reduce accordingly . . .'.

This observation is very pertinent and is in line with a discussion in Chapter 1 about the difference between a holistic and a reductionist approach to manufacturing problems and opportunities. The massive research effort directed towards generating better lot sizing algorithms and heuristics over the past number of years is the result of a reductionist fixation with this portion of the overall production and inventory management problem and a belief that by focusing on the consequences of the tradeoff between set-up and carrying costs, significant savings could be achieved. Few researchers have considered why set-up costs are high. They have simply accepted that they are high and went on from that assumption. A holistic approach would have questioned the high set-up costs which presumably would have led to a drive to reduce them. This is the approach that seems to have been adapted in Japan and it will be discussed in more detail when we look at the JIT (Just in Time) approach in Part III.

A Delphi study reported by Benson *et al.* (1982) posed the following question. What do you believe will be the most commonly used lot sizing technique by the end of 1990?

The answers indicated that sophisticated lot sizing techniques will not be used in the foreseeable future. 'Lot for lot, least total cost, and period order quantity were selected as the dominant techniques for 1990 by the majority of respondents'.

9.10 Conclusion

In this chapter, lot sizing techniques for use in MRP systems were reviewed. Practice seems to support the use of simple and *ad hoc* techniques. The last word is left to Burbidge (1985b) who said: 'In many ways perhaps the simplest argument against the EBQ is that it solves the wrong problem. The EBQ theory states that if set-up times are long one should make it in large batches too spread the set-up costs. A better argument is that if set-up times are long they should be reduced'.

This chapter concludes the explanation of MRP technology. In Chapter 10, various criticisms of the approach and the status of MRP II as a production management paradigm will be discussed.

CHAPTER TEN

The status of MRP/MRP II as a paradigm for PMS

10.1 Introduction

In this chapter, the status of MRP as a production management paradigm is discussed (see Harhen 1988). Five perspectives will be adopted in this review.

(1) The practice of MRP/MRP II.
(2) The reasons for failure of MRP systems.
(3) Current research and development related to MRP/MRP II.
(4) Criticisms of MRP.
(5) The state of the philosophical debate underlying MRP/MRP II.

10.2 The state of practice of MRP/MRP II

There is a significant divergence between what is available in MRP software packages and what has turned out to be useful. Therefore, to assess MRP/MRP II and to understand what is practice, is not a question of understanding the range of functions and options that are embedded in MRP II software. If that was the question, then the answer would be that state-of-the-art MRP is represented by a bucketless and net change MRP II system, which puts appropriate emphasis on master planning and supports the other functions (capacity, PAC, etc.) adequately.

The question of what is the state of the practice of MRP II, in our view, relates primarily to understanding the effectiveness of MRP systems for the companies that use them.

The pragmatics of operating an MRP system, although originally presented by Orlicky, find strong expression in the works of Wight (1981). Wight proposed a classification scheme that seeks to rate how well companies operate their MRP systems. The scheme is quite simple and involves a series of 25 questions which relate to the technical capability of the MRP software package, the accuracy of supporting data, the volume of education that has been provided to the employees and the results achieved by using the system. MRP system use is rated between class A, which represents excellence, and class D, which represents a situation where the only people using the system are those in the MIS (Manufacturing Information Systems) department.

Among the criteria that measure effective use of MRP are the following:

- MRP should use planning buckets no larger than a week.
- The frequency of replanning should be weekly or more frequent.
- If people are effectively using the system to plan, then the shortage list should have been eliminated.
- Delivery performance is 95% or better for vendors, the manufacturing shop and the MPS (Master Production Schedule).
- Performance in at least two of the following three business goals has improved:
    - inventory,
    - productivity,
    - customer service.

The various surveys taken through the years indicate several problems with MRP system implementations:

- Only a very small percentage of users of MRP consider themselves to be successfully operating their MRP systems. Many systems are *installed*, as opposed to *implemented*, i.e. the formal system is not the real system. The APICS study (Anderson *et al.* (1982), Schroeder *et al.* (1981)) showed that only 9.5% of MRP users considered themselves to be class A MRP II users, while 61.3% considered themselves to be class C or class D MRP II users.

  The LaForge *et al.* study (1986), though more recent, had a much smaller sample size. They found the percentage of class A MRP II users was 25%. The authors did not conclude that the percentage of firms with class A MRP II had increased and chose instead to say that the results supported the general pattern of MRP performance found by Anderson *et al.* (1982).
- Master production scheduling is not computerized by MRP users as often as might be expected. The Anderson study found that 52.2% of MRP users had computerized their MPS.

The LaForge study (1986) found the percentage of MRP users with computerized MPS was 61%.

- Capacity requirements planning also has a relatively low utilization by MRP users. The Anderson study found that 37.7% of MRP users had computerized their capacity planning system. LaForge (1986) reported 42%.
- In relatively few cases is computerized production activity control implemented. Anderson *et al.* found that 30.5% had done so.

  LaForge (1986) reported 52%.

On the other hand, important performance improvements stemming from MRP II were reported. For example, in the Anderson *et al.* study, the following improvements were identified:

- Average inventory performance improved from 3.2 to 4.3 turns.
- Average delivery lead times fell from 71.4 to 58.9 days.
- Average delivery performance rose from 61.4% to 76.6%.
- The average number of expediters or **progress chasers** fell from 10.1 to 6.5.

The findings were also supported by LaForge *et al.* What emerges from these surveys is that MRP II has definitely brought some major benefits to those manufacturing companies who have implemented it and that the costs associated with the implementation were, on average, well repaid. Class A MRP represents in many ways the state-of-the-art of MRP II. However, many firms have still not achieved this desirable goal.

10.3 Reasons for failure of MRP installations

We have seen that there have been many disappointing MRP installations. Lawrence (1986) reports on a major UK based study of CAPM (Computer Aided Production Management) systems which showed that of 33 companies studied only 16 claimed to have successfully implemented systems – 'Users do not seem to understand fully the facilities that are being offered to them in increasingly complex CAPM systems. Production controllers do not, on the whole, use the sophisticated algorithms for production scheduling made available to them in CAPM packages.' Many authors have tried to understand the background to MRP success and, indeed, failure. It seems to be generally agreed that failure of an MRP installation can be traced to problems such as:

- Lack of top management commitment to the project.
- Lack of education in MRP for those who will have to use the system.

- Unrealistic master production schedules.
- Inaccurate data, particularly BOM data and inventory data.

10.3.1 Top management commitment

Commitment by top management is seen as *essential* to the success of any MRP installation. Undertaking the installation of an MRP system is a major decision for any manufacturing company. It has implications for many areas throughout the manufacturing organization, for engineering (in terms of the need for accurate and completely up-to-date bills of materials), for purchasing (in terms of generating accurate purchase lead times) and for the materials and production people (in terms of the discipline necessary to maintain accurate inventory data and working to the schedule).

Safizadeh and Raafat (1986), among others, point to the fact that there are formal and informal systems within a manufacturing environment – 'At the time of MRP implementation, a well established, *somewhat accurate*, informal system is confronted with the demands and requirements of a new formal system. The installation of MRP may foster improved operations or it may lead to resistance and disintegration'. As the authors point out, MRP is inevitably about trying to use accurate and timely data and rigorous procedures in the production and inventory management function. This often involves a *culture* change for a group of people, in particular shop floor supervisors/ managers and *progress chasers* who have evolved a relatively efficient and well-tried manual informal system of shortage lists and priority schemes.

Latham (1981) argues that MRP 'touches, in some way, all the functionaries in an organization from the chief stock clerk to the chief executive officer, and that within most manufacturing organizations, MRP threatens long established habits and prerogatives which are born out of necessity and informal systems'. Latham goes on to appeal to production and inventory management professionals to learn 'additional skills, skills in dealing with the human aspects of systems'.

It is clear that if the manufacturing organization is to gain all of the potential benefits of introducing MRP, management must accept the responsibility for creating the environment which is amenable to, indeed positive in its support for, the changes which MRP implementation involves. Clearly such a favourable environment cannot be created without the full and enthusiastic involvement of top management.

Perhaps successful installation is most likely to be achieved by allowing the formal MRP system and the relatively informal pre-MRP system to sit side by side over a short period. The thinking is that those who have worked the informal system have the opportunity to become involved gradually in the new MRP system, while not feeling overawed, or even threatened, by it. However, it requires capable, sensitive and well informed management to ensure that all those involved gradually adapt the new formal MRP system and work together to achieve its full potential.

10.3.2 Education in MRP thinking and operation

A key element in any MRP installation is to ensure that all personnel in the company who are likely to come into contact with the MRP system should have some MRP education. Given the nature of MRP, many people in the manufacturing plant are impacted by its introduction. Therefore, a comprehensive MRP education programme has to be initiated to ensure that the system is used to its full potential. This is not to say that each employee from the chief executive officer down has to be an MRP expert, rather each should have sufficient understanding of MRP principles and operation to work with the system as required.

Hinds (1982) argues that 'it is during . . . the education process, . . . that the success of MRP is often determined. . . . Education is the first key to successful MRP implementation'. He concludes that 'the MRP process begins with, and its success is determined by, the education process, the goals of which are to support corporate objectives, acquire technical MRP knowledge, and create an atmosphere of company-wide cooperation'.

10.3.3 The need for accurate data

The MRP procedure is deceptively simple. After all, what is involved but the calculation of net requirements from gross requirements, taking the overall stock position into account, and then using some lot sizing technique to generate firm orders? Unfortunately life is not so simple. Earlier in this book we listed some prerequisites for MRP analysis, items such as availability of inventory data, BOMs, master schedule data, etc.

Perhaps the greatest requirement of all for successful MRP installation and operation is discipline. This includes the discipline to maintain accurate stock records, the discipline to report accurately and in good time the completion of jobs and orders, and the discipline to report to the system every event that MRP should be aware of. If stocks are withdrawn from stores, then this fact should be notified to the system and the inventory status in the production database updated accordingly. Many successful MRP installations have padlocks on the doors to the stockroom.

To conclude this discussion on the failure of MRP installations it is worth referring again to a Delphi study of manufacturing systems conducted in the early 1980s and reported by Benson *et al.* (1982). Among the questions the respondents in the study were asked to consider was the following: 'When will two-thirds of the attempted first-time MRP installations be successful?' Quoting Benson *et al.* the answers were summarized as follows: 'The majority of respondents believed that two thirds of the attempted first time MRP installations would be successful by the early to mid 1980s. However, about 18% said that this would not occur until after 2000 or would never occur'. Among the dissenting comments were: 'requires a behaviour change which is never easy the first time', 'too little understanding and too little user involvement will be the rule' and 'people are the problem – experience and education must increase greatly before this will occur'. Our view is that education and commitment to MRP success are important prerequisites for successful installation and operation of an MRP system.

10.4 Current MRP research and development

The main thrust of current development of MRP is the continued application of software engineering techniques to improve the MRP II system in terms of its user interface, its interconnectivity with other systems, data management and the provision of low cost delivery systems. None of these developments have changed the basic MRP procedure.

At the user interface level, recent developments have concerned user access to, and presentation of, information. Linking spreadsheets to MRP financial information is a developed capability in several MRP systems. The layering of decision support tools, such as fourth generation languages, on top of the MRP II database is also becoming commonplace. This is facilitated by the fact that today many MRP systems implement their data storage in database management systems, as opposed to earlier, less sophisticated, file based systems. Some user interface research has explored the provision of significant graphical display in MRP system outputs, particularly where concerned with capacity management and WIP management. While technically feasible, this has, until now, been restricted by the availability of low cost terminals with adequate graphics capability. Fortunately, this cost constraint seems to be softening.

The problem of interconnecting MRP II to other manufacturing information systems is currently attracting a great deal of research attention. It is part of the general CIM problem and today, at any manufacturing conference, one will find papers presenting case studies of how an MRP II system was linked to a CAD system, an automatic storage and retrieval system or a flexible manufacturing system. The appropriateness of various approaches to constructing these interconnections is not fully understood today, partly because the application of modular design in the large scale, multivendor environment that is CIM, is not itself understood. There are significant open questions as to how control systems that drive intelligent manufacturing systems, such as flexible manufacturing systems, will link with MRP II. Similar problems exist with bill of materials information, which appears in a company in many forms in both design engineering and manufacturing engineering systems.

Today MRP systems can be interconnected with such systems by customized interfaces, but this *interconnection* does not represent an *integration* in accordance with an understood and generally agreed architecture. Initial research in this direction is being focused on the developments of appropriate architectures and standards to cope with the problem of a large scale multivendor, multi-application environment. This is evidenced by activities around the higher levels of the Manufacturing Automation Protocol (MAP) specification (Kosmalski 1984), the work of the National Bureau of Standards in the USA (Simpson *et al.* 1984) and various efforts within the European Community's ESPRIT programme (Project 477, for instance (Actis-Dato *et al.* 1986, McCahill 1987)). One of the potential consequences of these efforts may be that MRP II will have to become more modular than is currently the case. Certain existing modules, such as bills of materials and production activity control, may have a separate existence at MRP II's interface with other environments, such as computer aided process planning, equipment control and material handling control.

Another development related to the interconnection of MRP II systems is Electronic Data Interchange (EDI). Emerging standards activity in the area of EDI will play an important role in facilitating the interconnection of the MRP system of one company with those of its suppliers. This has two purposes: to shorten the purchasing cycle and to transfer appropriate information for longer range material and capacity planning back to suppliers. This is currently being practised to a limited extent and the pressure to move to Just in Time manufacturing will accelerate this process.

On the factory floor, automatic data collection systems have been linked with the material tracking systems of MRP II systems, thus providing real-time access to WIP information. The interfaces are available off the shelf from some MRP II system suppliers.

MRP II packages that run on low end mini-computers and micro-computers have recently become available at relatively low cost. These may represent viable approaches for the small firm. Nevertheless, the dominant research and development effort in MRP II has been towards enhancing the functionality of systems to cope with the needs of large scale manufacturing enterprises. Little has been written about the suitability and acceptance of MRP as an appropriate approach for the very small firm.

Other research has focused on issues such as MRP/JIT/OPT comparisons with limited use of simulation models to test theories. Hall (1981) presented an extensive comparison of MRP and Kanban. In particular, he documents an ongoing effort to apply both the MRP and Kanban concepts in conjunction, namely **synchro-MRP**. Synchro cards, similar to Kanban cards, are used to control the flow of material in the repetitive flow areas in the factory. The synchro cards are generated from the mature portion of the master production schedule/final assembly schedule. These synchro cards can be used in the same way as Kanban cards to control the amount of WIP in the factory and so identify flow problems in the process. The supporting MRP system generates two types of manufacturing orders – a typical job shop order and a flow order. Parts are closed against the flow order as completed but parts are not produced or conveyed in this order size and are, instead, produced according to a more detailed schedule. Synchro-MRP requires the same rigorous engineering of the process as is the case with Kanban.

Attention has also focused on how to achieve some just in time goals by using MRP II (Sonnerbrug 1985). Approaches include the use of blow-through BOMs, recognition of operation overlap in shop routings, use of daily rather than weekly schedules and devoting specific attention to lot sizing and lead time rules. One emerging view is that MRP provides the planning framework within which the JIT execution system will operate.

This is, however, countered by those who say that MRP II really has nothing useful to say about lead times and set-up, and that the effective manufacturing systems of the future will be totally Just in Time. The truth of the matter is probably somewhere in between. JIT is better than MRP II where it can be applied. However, it may not always be applicable. Set-up and lead times may be economically irreducible in many situations.

10.5 Important criticisms of the MRP approach

Burbidge (1985a) identifies some of the key deficiencies within the MRP paradigm. In answer to a question put by one of the authors, Burbidge pointed to the long planning horizon normally associated with the master schedule activity in MRP and the consequent errors due to the inability of master schedulers to make accurate forecasts of demand towards the latter end of the planning horizon. The long planning horizon arises because of what Burbidge considers the inflated lead times associated with the MRP approach – 'MRP systems break the bill of materials into main, sub, sub-sub, . . . etc. assembly and fabrication stages, estimate lead times for each stage and add them together to establish lead times for ordering. This inflates lead times and stocks'.

As will be seen in Part IV, when the OPT approach is discussed, this determination of lead times is seen by many people as a fundamental weakness in the MRP approach to production and inventory management. From an integration or CIM point of view, the MRP philosophy seems to accept long lead times as fixed and given and does not seek to reduce them. This is not to say that MRP users are unaware of the consequences of long queue times and lead times. Rather, MRP II is concerned only with estimating and using lead times and not with reducing them *per se*.

In fact, the lead time used in MRP offsetting is the planned or expected lead time and this represents no more than an estimate of the time it takes for an individual batch to go through the system. The actual lead time will depend on the load on the manufacturing shop floor and the priority assigned to a given batch.

One of PAC's functions is to *close the loop* between the MRP planning system and the manufacturing shop floor. One element of this *closing the loop* process is the feedback of actual lead times to the MRP system. This data can then be used to establish the validity of the lead times in use and to signal the need for a change if necessary.

Thus the concept of a *planned lead time* involves a major assumption. MRP assumes that the lead time is known in advance of the schedule and is, furthermore, independent of the batch size. This is *true* if the largest component of lead time is the queuing time. However, this large queuing time reflects an inefficient flow of work through the manufacturing shop and perhaps a process rather than a product based layout. Furthermore it is an *average* queuing time whose value for a particular batch or job will vary from week to week, reflecting the load on the manufacturing shop and, in particular, on the work centre in question. This notion of scheduling based on *known* lead times makes MRP essentially capacity insensitive and is one of the fundamental criticisms of MRP which the OPT approach to production control (see Part IV) seeks to address.

Another difficulty of MRP is its treatment of batch or lot sizes. The APICS dictionary (Wallace 1980) defines a lot size as 'the amount of a particular item that is ordered from the plant or vendor'. However, there are many other possible interpretations of the term lot or batch. Burbidge (1985b) identifies four possible definitions of a batch or lot:

(1) 'Run Quantity (RQ) is the quantity of parts run off at a work centre before changing to make some other part.
(2) Transfer Quantity (TQ) is the quantity of parts transferred as a batch between the work centres for successive operations.
(3) Set-up Quantity (SQ) is the quantity of parts, not necessarily all the same, which are produced at a work centre between changes in the tooling set-up.
(4) Order Quantity (OQ) is the quantity of parts authorized for manufacture or purchase, by the issue of a written order.'

The batch or lot in MRP is clearly what Burbidge terms the order quantity. The notion of a transfer quantity is an important one and has important ramifications for the lead time calculation. In fact, as will be seen later, the transfer batch plays an important part in JIT and OPT thinking. Let us take a simple example to illustrate the importance of separating the order quantity from the transfer quantity.

Assume we have an order quantity of 50 for a component which requires two machining operations, each of which involves ten minutes machine time per operation. Let us further assume that in this case there is no queuing time and that the transfer time between operations is zero.

- **Case 1** Transfer quantity = Order quantity = 50
  We can only transfer the complete batch from machine 1 to machine 2. We cannot transfer any component from the first machine to the second machine until all 50 have been machined. The time to complete the 50 items is ($50 \times 10 \times 2$) or 1000 minutes, or 16 hours and 40 minutes as in Figure 10.1.

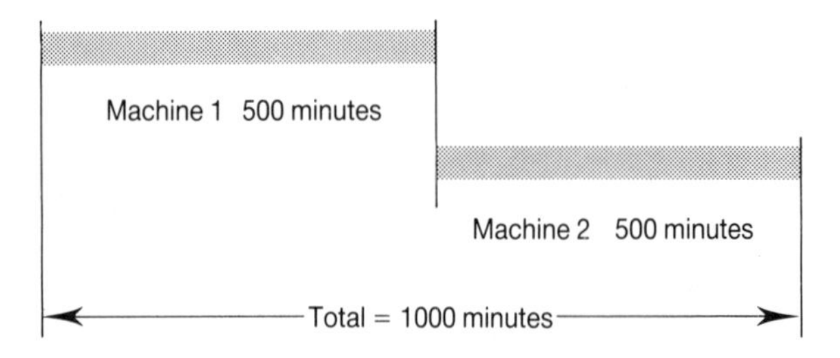

Figure 10.1 Transfer quantity equal to order quantity.

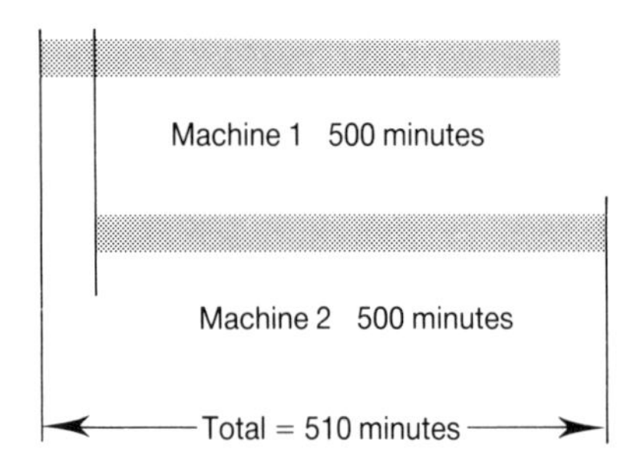

Figure 10.2 Transfer quantity equal to one.

- **Case 2** Order quantity = 50. Transfer quantity = 1
  As soon as a component is finished on machine 1 it can be transferred to the second operation on machine 2. The time to complete 50 items is 510 minutes, or 8 hours and 30 minutes as in Figure 10.2.

This is, of course, an extreme case but, nevertheless, it illustrates the point. The only concession the MRP approach makes to this type of thinking is the **split lot**. The split lot is defined by the APICS Dictionary as 'a manufacturing order quantity that has been divided into two or more smaller quantities usually after the order is in process. Lots are sometimes split so that a portion of the lot can be moved through manufacturing faster. This portion is called the send-ahead'.

Batch splitting is not normally encouraged within MRP and it is seen as a facility which may be used only to progress a late and/or urgent job.

Why MRP assumes that there is only one batch is quite difficult to explain. Perhaps it goes back to the thinking behind batch production, that is, process based layouts, long lead times and long set-up times.

10.6 The philosophical debate concerning MRP/MRP II

To understand what is the status of MRP/MRP II as a production management paradigm also requires some consideration of the state of the philosophy of MRP users. Two major influences have emerged to put pressure on the MRP II paradigm. There is the decision science concern of how MRP II compares with emerging popular alternative philosophies, such as just in time and OPT. There is also the software engineering concern of where the limits of an MRP II system should be and how MRP integrates with proliferating factory automation and the controllers that supervise it. As already stated, the road to CIM seems to involve some redefinition of what functions remain within or move outside the MRP II system, particularly in the areas of BOM management and PAC.

An essential and core proposal of the MRP paradigm is that the production management system should use a very simple technique, so that people may understand what decisions are being made by the computer and what human interventions are appropriate. Once people understand and are motivated then they can be expected to assume responsibility for ensuring that the large amount of raw data processed by the system is accurate and that the recommendations made by the system are valid.

Therefore, since finite loading algorithms are necessarily heuristic and probably difficult for the layman to understand, they have been frowned upon by the MRP community. MRP II is seen in the role of an infinite loading/decision support tool, wherein the users develop the schedule particularly using the *what if* support provided by the MPS system. MRP II is also a hierarchical scheduling system since scheduling decisions are made at three levels of aggregation – the MPS level, the MRP level and the PAC level. This prompts the question of whether decision support/infinite loading is the paradigm that is appropriate across each of the levels.

Hierarchical scheduling systems seem to be a way for the future. To the extent that manufacturing processes become more highly engineered, automated and predictable, and the people who operate these processes become more highly skilled, then it seems less reasonable to adopt an infinite loading/decision support strategy for scheduling manufacturing, particularly at PAC/MRP level, regardless of the *keep the responsibility with the people* argument. On the other hand, to the extent that manufacturing processes remain people intensive and less predictable, then finite loading becomes a very dubious proposition. Moreover, it is unlikely in any case that at the master planning level, any approach other than a decision support will ever be acceptable. The claim that MRP II will be made obsolete by OPT (Fox 1985) with its partitioned forward-finite/backward-infinite heuristic seems unlikely to be fulfilled. Nevertheless, it seems likely that both finite and infinite paradigms will survive and in the future we will continue to see emerging hybrid approaches.

The debate around JIT seems to have reached some tentative understanding around some of the issues. MRP users can and have learned a great deal from Kanban about the evils of work in progress, for example. This has stimulated great attention to cycle times, throughout the whole manufacturing process. It seems that in repetitive manufacturing systems, MRP performance is inferior to a well engineered process using the JIT/Kanban approach. However, in non-repetitive situations, such as job shop and small batch production, full Kanban is difficult to implement and MRP remains a very workable solution. In between these situations there is probably much room for hybrid applications of both techniques.

In evaluating scheduling systems it is important to distinguish between the scheduling technique in itself and the technique as applied to a real system. The choice of technique may not be overly important compared with the need to improve the management and engineering process that supports the application of that technique. The point was well made recently by Galvin (1986) when he stated that:

> 'Apparently the techniques employed are not such a dominant factor as we have been led to believe . . . The cohesive power of a successful system would appear to be due to the readiness and concerted efforts of all functional groups to achieve a common plan. If you haven't got a common plan, then you haven't got a system'.

10.7 Conclusion

This chapter has attempted to discuss the status of MRP/MRP II as a production management paradigm. Practice, research and philosophy have all been considered. In conclusion, MRP/MRP II is a viable approach to production management, with a proven track record. While MRP II will continue to be widely applied in its present form it may, nevertheless, be subject in the future to radical modularization in order to re-emerge in new hybrid production management environments.

References

Actis-Dato, M., Erhet, O. and Barta, G. 1986. 'Control systems for integrated manufacturing: the CAM solution', in *ESPRIT 86: Status Report of Ongoing Work*, edited by the Commission of the European Communities. Amsterdam: North-Holland.

Anderson, J., Schroeder, R., Tupy, S. and White, E. 1982. 'Materials requirements planning systems: the state-of-the-art', *Production and Inventory Management*, **23**(4), 51–67.

Benson, P., Hill, A. and Hoffman, T. 1982. 'Manufacturing systems of the future: a Delphi study', *Production and Inventory Management* **23**(3), 87–106.

Berry, W. 1972. 'Lot sizing techniques for requirements planning systems: a framework for analysis', *Production and Inventory Management*, **13**(2).

Berry, W., Vollman, T. and Whybark, D. 1979. 'Master production scheduling, principles and practice', Washington DC: American Production and Inventory Control Society.

Burbidge, J. 1985a. 'Automated production control', in *Modelling Production Management Systems*, edited by P. Falster and R. Mazumber. Amsterdam: North-Holland.

Burbidge, J. 1985b. 'Production planning and control: a personal philosophy', in *Proceedings of IFIP WG 5.7 Working Conference on Decentralized Production Management Systems*, Munich, FRG, March.

DeBodt, M., Van Wassenhove, L. 1983. 'Lot sizes and safety stocks in MRP', *Production and Inventory Management*, **24**(1).

DeMatteis, J. and Mendoza, A. 1968. 'An economic lot sizing technique', *IBM Systems Journal*, **7**(1), 30–46.

Fordyce, J. and Webster, F. 1984. 'The Wagner-Whitin algorithm made simple', *Production and Inventory Management*, **25**(2).

Fox, R. 1985. 'Build your own OPT', in *APICS 28th Annual International Conference Proceedings*, Falls Church, VA: American Production and Inventory Control Society, 568–572.

Gaither, N. 1981. 'A near optimal lot sizing model for MRP systems', *Production and Inventory Management*, **22**(4).

Gaither, N. 1983. 'An improved lot sizing model for MRP systems', *Production and Inventory Management*, **24**(3).

Galvin, P. 1986. 'Visions and realities: MRP as system', *Production and Inventory Management*, **27**(3).

Groff, G. 1979. 'A lot sizing rule for time phased component demand', *Production and Inventory Management*, **20**(1).

Hall, R. 1981. *Driving the Productivity Machine: Production Planning and Control in Japan*, Falls Church, VA: American Production and Inventory Control Society.

Harhen, J. 1988. 'The-state-of-the-art of MRP/MRP II', in *Computer Aided Production Management: The State-of-the-Art*, edited by A. Rolstadas. FRG: Springer-Verlag (in press).

Hinds, S. 1982. 'The spirit of materials requirements planning', *Production and Inventory Management*, **23**(4), 35–50.

Ho, C., Carter, P., Melnyk, S. and Narasimhan, R. 1986. 'Quantity versus timing change in open order: a critical evaluation', *Production and Inventory Management*, **27**(1), 122–138.

Hoyt, J. 1983. 'Determining dynamic lead times for manufactured parts in a job shop', in *Computers in Manufacturing Execution and Control Systems*. New Jersey, USA: Auerbach Publishers.

Kochhar, A. 1979. *Development of Computer Based Production Systems*. London: Edward Arnold.

Kosmalski, D. 1984. *Manufacturing Automation Protocol Specification*. Warren, Michigan: General Motors Corporation.

LaForge, R. and Sturr, V. 1986. 'MRP practices in a random sample of manufacturing firms', *Production and Inventory Management*, **28**(3), 129–137.

Latham, D. 1981. 'Are you among MRP's walking wounded?', *Production and Inventory Management*, **22**(3), 33–41.

Lawrence, A. 1986. 'Are CAPM systems just too complex?', *Industrial Computing*, September, p. 5.

Mather, H. 1985. 'Dynamic lot sizing for MRP: help or hindrance', *Production and Inventory Management*, **26**(2).

McCahill, J. 1987. 'Towards an application generator for PAC in a CIM environment', in *ESPIRIT 87: Status Report of Ongoing Work*, edited by the Commission of the European Communities. Amsterdam: North-Holland.

Orlicky, J. 1975a. *Materials Requirements Planning: The New Way of Life in Production and Inventory Management*. New York: McGraw-Hill.

Orlicky, J. 1975b. 'A note on exercises in sterility', *Production and Inventory Management*, **16**(3), 90–91.

Orlicky, J. 1976. 'Rescheduling with tomorrow's MRP system', *Production and Inventory Management*, **17**(2) 38–47.

Safizadeh, M. and Raafat, F. 1986. 'Formal/informal systems and MRP implementation', *Production and Inventory Management*, **27**(1).

Schroeder, R., Anderson, J., Tupy, S. and White, E. 1981. 'A study of MRP benefits and costs', *Journal of Operations Management*, **2**(1), 1–9.

Silver, E. and Meal, H. 1973. 'A heuristic for selecting lot size quantities for the case of a deterministic time varying demand rate and discrete opportunities for replenishment', *Production and Inventory Management*, **14**(2).

Simpson, J., Hocken, R. and Albus, J. 1984. 'The automated manufacturing research facility of the National Bureau of Standards', *Journal of Manufacturing Systems*, **1**(1), 17–32.

Sonnerbrug, R. 1985. 'The MRP and Just in Time marriage', in *APICS 26th Annual International Conference Proceedings*. Falls Church, VA: American Production and Inventory Control Society, 683–687.

St. John, R. 1984. 'The evils of lot sizing in MRP', *Production and Inventory Management*, **25**(4).

Vollman, T., Berry, W. and Whybark, D. 1984. *Manufacturing Planning and Control Systems*. Homewood, Illinois: Dow Jones Irwin.

Wagner, H. and Whitin, T. 1958. 'Dynamic version of the economic lot size model', *Management Science*, **5**(1), 89–96.

Wallace, T. 1980. *APICS Dictionary*, 4th Edition, Washington, DC: American Production and Inventory Control Society.

Wemmerlov, U. 1979. 'Design factors in MRP systems: a limited survey', *Production and Inventory Management*, **20**(4), 15–34.

Wight, O. 1981. *MRP II: Unlocking America's Productivity Potential*. Boston, MA: CBI Publishing.

PART III

Just in time

Overview

Just in Time (JIT) production has attracted the attention of management in the 1980s. Western industrial managers, aware of the success of their Japanese counterparts, now believe that a commitment to achieving just in time in manufacturing is essential in order to compete on worldwide markets. In Part III, the just in time concept, its influence on production management and the movement towards computer integrated manufacturing are discussed.

Our basic contention is that the Kanban card system is greatly over-emphasized in the literature on JIT, at the expense of the JIT approach and JIT manufacturing techniques. In particular, we feel that JIT can be applied to good effect in all types of discrete parts manufacturing systems while Kanban can only be used in repetitive manufacturing systems.

The structure of Part III is very different to that of Part II. The discussion of the MRP approach in Part II started by looking at the mechanics of MRP and went on to discuss the MRP paradigm much later. In Part III the opposite approach is adopted, that is, Chapter 11 starts with the thinking underlying JIT and moves towards the mechanics, i.e. the techniques for reducing set-up times, Kanban, etc. in the later chapters. This is in line with the view expressed earlier, i.e. that up to now, not enough emphasis has been placed on the fundamental concepts of JIT.

Part III is structured as follows. In Chapter 11 the fundamental concepts behind Just in Time are introduced and the philosophy inherent in the approach is outlined. Chapter 12 describes how the manufacturing environment is designed and production planned in order to set up a context within which just in time production can be achieved. Finally, Chapter 13 attempts to outline how control of production is maintained in a just in time system using Kanban which, in our view, is simply a manual shop floor control system.

CHAPTER ELEVEN

The just in time approach

11.1 Introduction

The success of Japanese firms in the international marketplace has generated interest among many Western companies as to how this success was achieved. Many claim that the keystone of the Japanese success in manufacturing is Just in Time (JIT). Just in time is a manufacturing philosophy with a very simple goal, i.e. produce the required items, at the required quality and in the required quantities, at the precise time they are required. JIT has been described by Schonberger (1984) as a production system which replaces 'complexity with simplicity in manufacturing management'.

The technical literature is filled with articles and case studies exhorting the JIT system and encouraging management to implement it. For example, Cortes-Comerer (1986) claims that 'At Hewlett-Packard Company's plant in Cupertino, California, . . . the time it takes to assemble a set of 31 printed circuit boards has been slashed from 15 days in 1982 to 11.3 hours in 1986. During the same period the inventory of circuit board work in process was cut from $670 000 to $20 000 and the number of back orders was reduced from an average of 200 to 2. . . . The secret of this success story? Just in time manufacturing or JIT . . .' Many companies are reported to have had similar experiences with JIT systems.

The JIT system arose initially in the Toyota automotive plants in Japan in the early 1960s and is currently being used in a variety of industries, including automotive, aerospace, machine tools, computer and telecommunications manufacturing.

Just in time can be viewed from three perspectives, all of which must be considered in order to achieve JIT. This is illustrated in Figure 11.1.

The shop floor control system for JIT is the most visible manifestation of the JIT approach because of its use of Kanban cards. The Kanban technique controls the initiation of production and the flow of material with the aim of getting exactly the right quantity of items (components, sub-assemblies or purchased parts) at exactly the right place at precisely the right time. Its use is described in detail in Chapter 13.

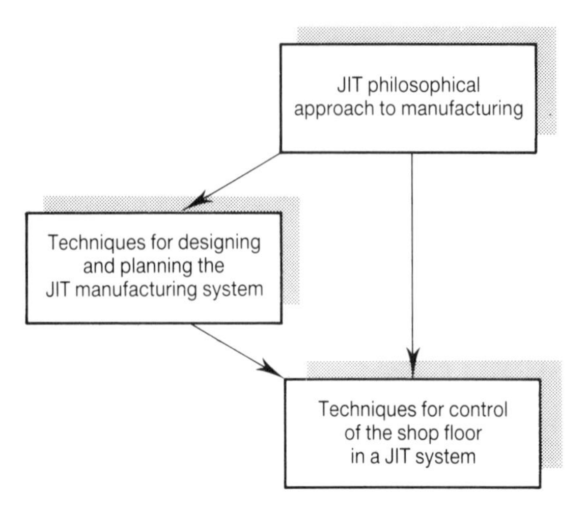

Figure 11.1 JIT approach.

Underlying the use of Kanban is the prior application of an array of techniques to the products and manufacturing processes, in order to ensure that the application of Kanban is feasible. The techniques involve the design of the manufacturing system in its broadest sense, addressing issues of marketing, sales, product design, process engineering, quality engineering, plant layout and production management, in order to facilitate JIT production using the Kanban system.

The third and most fundamental level is the JIT philosophy of manufacturing on which JIT execution and the design and planning of the JIT manufacturing system are premised. This is frequently the least understood aspect of JIT but in many ways it is the most important. The JIT philosophy is a set of fundamental manufacturing strategies which, when implemented, provide the basis for the JIT system and facilitate the use of the Kanban system.

This chapter will concentrate on reviewing the JIT philosophy, looking first at the goals of JIT and then discussing the key ideas making up the approach. Chapters 12 and 13 will examine the design and planning of the manufacturing system and finally the Kanban system.

11.2 The goals of the JIT approach

The JIT approach involves a continuous commitment to the pursuit of excellence in all phases of manufacturing systems design and operation.

JIT seeks to design a manufacturing system for efficient production of 100% good units. It seeks to produce only the required items, at the required time and in the required quantities. This is probably the simplest statement of the JIT approach to manufacturing.

To be more specific, JIT seeks to achieve the following goals (Edwards 1983):

- zero defects
- zero set-up time
- zero inventories
- zero handling
- zero breakdowns
- zero lead time
- lot size of one.

There are two aspects of the set of goals listed above which are worth noting:

(1) In the minds of many manufacturing or industrial engineers trained in the *Western* approach to manufacturing systems design and operation, these goals seem very ambitious, if not unattainable.

(2) The attempt to consider all of these goals simultaneously is unusual in the context of the traditional approach to manufacturing systems. As pointed out earlier in Chapter 2, the traditional approach to manufacturing has been reductionist, which involves consideration of well defined aspects of the overall manufacturing problem – in fact, separate sub-problems – which are tackled and *solved* as separate problems. This approach has led to the proliferation of specialists in the various manufacturing functions, with a resulting absence of any generalist to consider the whole of the manufacturing system. The JIT approach can clearly be characterized as holistic, at least in terms of the range of goals it sets for itself.

11.2.1 Zero defects

In traditional manufacturing management the goal of *zero defects* is rarely considered. In fact, quality people have traditionally thought and planned in terms of LTPD (Lot Tolerance Per Cent Defective) and AQLs (Acceptable Quality Levels). The emphasis in these traditional systems is likely to be on inspection systems, control charts and *acceptable* quality levels for the items produced. The underlying assumption seems to be the belief that a certain level of unacceptable product is unavoidable and that the emphasis should be on reaching an attainable or acceptable level of conformity to specification and to customer expectation. This contrasts with the JIT approach, which aims to eliminate once and for all the causes of defects, and so engenders an attitude of seeking to achieve excellence at all stages in the manufacturing process.

11.2.2 Zero inventories

In traditional manufacturing thinking, inventories, including Work in Progress (WIP) and the contents of finished goods stores, are seen as assets in the sense that they represent added value which has been accumulated in the system. From the perspective of the shop floor supervisor, inventories are also *good* in that they represent a build-up of work available in the supervisor's department. Furthermore, at the end of the week, the difference between the starting (i.e. beginning of the week) inventory and the inventory on hand represents a part of the *value added* during the week and tends to indicate increased efficiency for the department in question.

Inventories are seen, in many cases, as a buffer against uncertain suppliers in the case of raw materials and bought-in items. Outside suppliers are *distrusted* and the thinking is almost to assume that they may not deliver on time and hence the buffers – as *insurance* against uncertain availability of work by shop floor supervisors and as a buffer against an unexpected customer order by the marketing and sales function.

Moreover, the quest for manufacturing efficiency where efficiency is measured in terms of the utilization of equipment within shop floor departments, encourages shop floor supervisors and managers to keep individual machines and work centres busy continuously, producing items which are often not mandated by current orders and are perhaps required to meet future, as yet unannounced, demand. The expensive work centres, those with the highest cost or capital/overhead recovery per hour, tend to be singled out for special attention and a *good* supervisor works hard to keep such machines busy. Ironically, such expensive work centres are likely to be the most productive in the plant and therefore capable of generating huge amounts of inventory. One can see that the emphasis on *recovering* overhead by keeping machines busy stems from a reductionist approach to manufacturing since the overall cost of such behaviour is ignored.

Clearly, the effect of this emphasis on recovering overhead is to build-up inventories to levels higher than they might otherwise be. Occasionally pressure may be exerted by senior management to reduce the levels of stock in the plant, but this will be buried as part of an overall drive to increase productivity and manufacturing efficiency, and so we get bound again by the *utilization of equipment* goal. In addition, high inventories are traditionally seen as the responsibility of the materials management function in manufacturing organizations – particularly the production controllers – while efficiency/utilization are traditionally the responsibility of shop floor supervisors and management.

The accountants see inventory as an asset to be recorded on the balance sheet. Senior management may not have encouraged the build-up of large stocks but one feels that faced with a choice between low plant utilization with low stocks on the one hand and high plant utilization and high stocks on the other, they tend to choose the latter.

All of this contrasts with the JIT view that inventory is *evil* and that inventory is evidence of poor design, poor coordination and poor operation of the manufacturing system.

11.2.3 Zero set-up time

The concepts of zero set-up time and a lot size of one are interrelated. If the set-up times are approaching zero, then this implies that there is no advantage to producing in batches. As seen earlier, the thinking behind the economic order quantity/economic batch quantity approach is to minimize the total cost of inventory by effecting a trade-off between the costs of carrying stock and the cost of set-ups. Very large batches imply high inventory costs. Very small batches result in correspondingly lower inventory costs but involve a larger number of set-ups and, consequently, larger set-up costs. However, if set-up times and costs are zero, then the ultimate small batch, namely the batch of one, is economic. The consequences of a lot size of one are of enormous benefit from an inventory and overall manufacturing performance perspective.

11.2.4 Zero lead time

An equally important result of small lots is the effect they have on flexibility. Small lots combine with the resultant very short lead times to greatly increase the flexibility of the manufacturing system. It was seen in our discussion on the length of the planning horizon for master scheduling in MRP systems that the planning horizon must be at least as long as the longest cumulative product lead time. Long planning lead times force the manufacturing system to rely on forecasts and to commit to manufacturing product prior to, and in anticipation of, customer orders. Furthermore, if a batch of product has already progressed some way through a series of manufacturing processes, it is difficult, if not impossible, to modify the batch size (if so needed) by short term fluctuations in market demand patterns. Small lot sizes, combined with short lead times, mean that the manufacturing system is not committed to a particular production program over a long period and can more readily adapt to short term fluctuations in market demand.

To approach zero lead time, the products, the manufacturing system and the production processes must be so designed as to facilitate rapid throughput of orders. Traditional approaches tended to treat product and process design separately. The JIT philosophy takes a holistic approach and recognizes the interdependence between these activities.

The importance of the *zero lead time* goal cannot be overstated when considering the demands placed by the market on manufacturers to respond quickly to orders for a diversity of products (see Chapter 1). While zero lead time is impossible, a manufacturing system that pursues such an ideal objective and constantly strives to reduce the lead times for products to the absolute minimum, will tend to operate with greater flexibility than its competitors.

11.2.5 Zero parts handling

Manufacturing and assembly operations frequently include a large number of non-value adding activities. Taking assembly operations as an example, many assembly tasks can be viewed as a combination of the following operations:

- component feeding,
- component handling,
- parts mating,
- parts inspection,
- special operations.

Operations such as *component feeding* and *component handling* are non-value adding operations. If components and assemblies could be designed to minimize feeding and if manufacturing systems could be designed to minimize handling, significant reductions in assembly problems and assembly times could be achieved. (See Boothroyd and Dewhurst (1982, 1983) and Browne *et al.* (1986) for some design guidelines on how to achieve this.)

As Boothroyd and Dewhurst (1983) stated, 'design is the first stage of manufacturing. It is here that the manufacturing costs are largely determined. In addition, the assembly process is usually the single most important process contributing to both manufacturing costs and labour requirements'.

As will be seen later, the product based manufacturing layout is preferred to the traditional process based layout. One reason for this is that product based layout results in much simpler patterns of material flow through the plant and, consequently, considerably reduces the planning and materials handling effort.

11.3 Key elements in the JIT approach

Given the goals discussed above the essential elements of the JIT philosophy for product and manufacturing system design can be determined. These important elements are:

- An intelligent match of the market demand with product design in an era of greatly reduced product life cycles and with the early consideration of manufacturing problems at the product design stage.

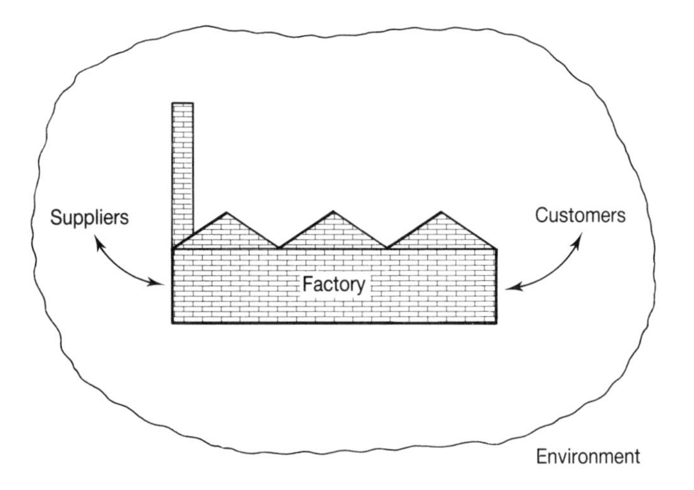

Figure 11.2 Plant environment from JIT perspective.

- The definition of product families based on a number of important manufacturing goals and the design of manufacturing systems to facilitate flow based production of these families where possible.
- The establishment of relationships with suppliers to achieve just in time deliveries of raw materials and purchased components.

These three elements can be viewed as part of an overall approach to manufacturing which sees the factory sitting within an environment as shown in Figure 11.2. The front end involves the factory and its relationship with its customers in the marketplace and the back end is the relationship of the factory with its suppliers. The fact that the JIT approach considers the total manufacturing picture is not surprising in view of the wide range of goals that JIT seeks to address and which have been outlined above.

The importance of this approach to manufacturing, i.e. not restricting attention to the *internals* of the factory, cannot be overstressed. Let us return for a moment to our discussion in Chapter 3 on the distinction between computer integrated business and computer integrated manufacturing.

In Chapter 3 it was pointed out that manufacturing lead time is the time taken from receipt of the order in the plant to the completion of the order and its availability for delivery to the customer. CIM, it was noted, focuses on that part of the business organization whose goal is to reduce the manufacturing lead time, increase the quality and reduce the cost of the product. At the front end of the business it takes a certain time to process the order from the customer and to acquire the necessary raw material and purchased items from vendors. Similarly, it takes some time to assemble the finished products and to dispatch the completed order to the customer.

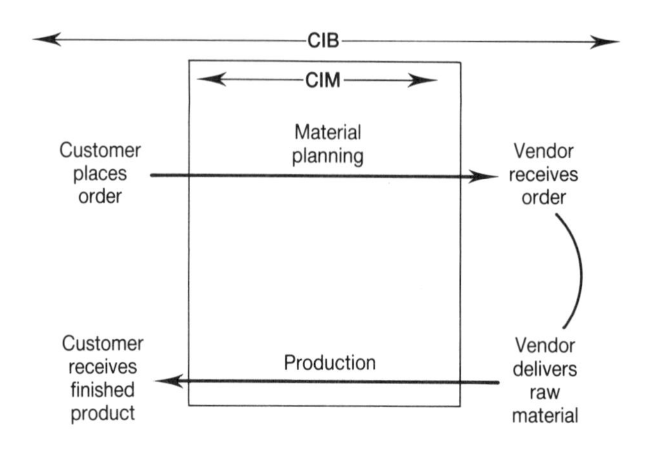

Figure 11.3 Time between when customer orders and receives product.

Figure 11.3, a replica of Figure 3.8, depicts this situation and suggests that, in many industries, the times for the processing of customer orders, the acquisition of the raw material and the distribution of the finished product are larger than the manufacturing lead time. This is so for a number of reasons, partially because of problems associated with product diversity and semi-customized products. For example, actual understanding of customer requirements and the conversion of this understanding into requirements for recognizable part numbers in the manufacturing database can take a long time and involve sales and technical sales support people.

As Figure 11.3 suggests, CIM is concerned primarily with activity during the manufacturing lead time, whereas computer integrated business is concerned with all aspects of the relationship with the customer, from receipt of initial order to dispatch of product and, perhaps subsequently, to maintain a relationship with the customer for maintenance and product update purposes. Reducing the manufacturing lead time is a key goal of modern manufacturing. Through the use of computer technology, great opportunities exist to reduce the time taken from initial contact with the customer through receipt of the order and onto the dispatch of the finished goods. This process of integrating the total business is termed CIB or computer integrated business.

In our view, the JIT approach to manufacturing incorporates a business perspective, as distinct from a narrow or strictly manufacturing (i.e. inside the four walls of the plant) perspective. The three key elements of the JIT approach identified above are a result of this business perspective and each of them will now be discussed in turn.

11.3.1 A match of product design to market demand

Chapter 1 reviewed the changed environment of manufacturing, focusing on greatly increased product diversity and greatly reduced product life cycles as important factors in this new environment. The heightened expectations of today's consumers, who demand considerable choice in the configuration of options, was also alluded to – the automotive market being an important example of this trend.

Of course, even with today's sophisticated and versatile manufacturing technology, companies cannot provide *customized* products at an economic price to the mass market. What is required is that industry interpret the wishes of the marketplace and, in a sense, direct the market in a manner that allows it (the industry) to respond to the market effectively. This involves designing a range of products which anticipate the market requirement and include sufficient variety to meet consumers' expectations and which can be manufactured and delivered to the market at a price which the market is willing and able to pay.

To achieve this objective it is necessary to design products in a modular fashion. A large product range and a wide variety of product styles can result in high manufacturing and assembly costs due to the high cost of flexibility in manufacturing systems. In general terms, it is true that the greater the flexibility required, the more expensive will be the manufacturing system and, therefore, the products of that manufacturing system. (The effect of the new computer based manufacturing technology is simply to shift the cost curve in the sense that flexibility becomes relatively less expensive – however, it is still true that there is a cost premium associated with the provision of flexibility.) Thus, too broad a product range and variety of product styles will result in product which is too expensive for the market.

Modular product designs are achieved by rationalizing the product range, where possible, and by examining the commonality of components and sub-assemblies across the product range with a view to increasing it to the maximum level possible. Rationalization of the product range results in reduced production costs through less manufacturing set-ups, fewer items in stock, fewer component drawings, etc. These issues will be considered in more detail in Chapter 12 when the product design issue will be reviewed further.

The approach here is somewhat akin to that of Skinner (1974) and his concept of the **focused factory**. Skinner suggests that a new management approach is needed in industries where diverse products and markets require companies to manufacture a wide range of products in forms and quantities:

> '. . . One way to compete is to focus the entire manufacturing system on a limited task precisely defined by the company's competitive strategy and the realities of its technology and economics . . . Instead of permitting a whirling diversity of tasks and ingredients, top management applies a centripetal force,

which constantly pulls inward towards one central focus – the one key manufacturing task. The result is greater simplicity, lower costs, and a manufacturing and support organization that is directed towards successful competition.'

11.3.2 Product families and flow based manufacturing

A common approach to the identification of product families and the subsequent development of flow based manufacturing systems is Group Technology (GT). In JIT systems, the use of GT to define product families is important for a number of reasons. Firstly, group technology is used to aid the design process and to reduce unnecessary variety and duplication in product design. Secondly, group technology is used to define families of products and components which can be manufactured in well defined manufacturing cells. The effect of these manufacturing cells is to reorient production systems away from the process based layout and towards the product or flow based layout. As Hyer *et al.* (1982) point out, group technology leads to cell based manufacturing which 'promises shorter lead times, reduced work in progress and finished goods inventories, simplified production planning and control and increased job satisfaction'. Group technology was not originally conceived by the Japanese but its philosophy was adopted by them and drawn into the JIT approach to manufacturing. Those readers interested in a short discussion on the historical background to group technology are referred to Gallagher and Knight (1973).

Group technology, in a sense, creates the conditions necessary for JIT because, as Lewis (1986) points out, it results in:

- 'Control of the variety seen by the manufacturing system.
- Standardization of processing methods.
- Integration of processes'.

In group technology, components are grouped into families on the basis of similarity of such features as part shapes, part finishes, materials, tolerances and required manufacturing processes. Gallagher and Knight (1973) define group technology as 'a technique for identifying and bringing together related or similar components in a production process in order to take advantage of their similarities by making use of, for example, the inherent economies of flow – production methods'.

Group technology forms component families on the basis of the design or manufacturing attributes – sometimes both – of the components in question. A large number of classification systems have been developed including the Brisch system in the UK and the Opitz system in the Federal Republic of Germany. These systems allow the manufacturing systems analyst to code the components manufactured in a plant and to identify families of components which have similar processing requirements and which, consequently, can be manufactured in a group technology cell.

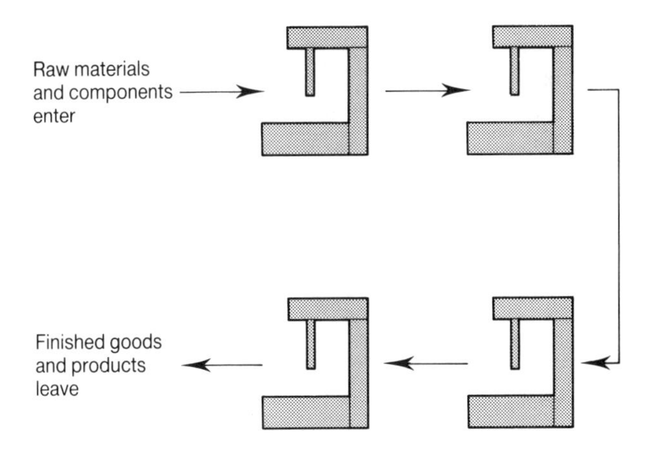

Figure 11.4 Cell layout.

The effect of these cells is to generate simplified material flow patterns in a plant, to allow responsibility and *ownership* for a component or group of components to rest with one group of operators and their supervisor.

These effects are best understood by considering the differences between a traditional, functional (process based) plant layout and a group technology (product based) layout. In the functional or process layout, machines are organized into groups by function. Thus, in a metal cutting machine shop the lathes would be grouped together, as would the the milling machines, the grinders, the drilling machines, etc. A departmental foreman would be responsible for a particular function or group of functions. Individual components would *visit* some, or maybe all, departments and thus would pass through a number of different supervisors' areas of responsibility. Operators and their supervisors would be responsible for different operations on each component but not for the resulting component or assembly itself. Given the variety of components associated with batch type production systems, the actual route individual batches take through the various departments or functions in the plant varies and the material flow system is thus complex. Furthermore, given this complex and virtually *random* material flow system, it is not easy to say, at any point in time, what progress has been made on individual batches.

The product or cell based layout, as shown in Figure 11.4, is clearly considerably simpler than the process layout. In fact, this simplicity is a hallmark of just in time systems and for many writers and researchers on manufacturing systems is a key characteristic of the system. (See, for example, Schonberger (1984)). As we shall see later this simplicity facilitates the use of a manual production activity control system, namely Kanban, on the shop floor itself.

Burbidge (1975) suggests that the process based layout is a poor basis for manufacturing efficiency and argues that process layout results in very long throughput times and does not facilitate the delegation of product responsibility down to the shop floor level.

A technique used to plan the change from a process to a product based plant organization is Production Flow Analysis (PFA) (see Burbidge (1963)). Production flow analysis is a technique based on the analysis of component route cards which specify the manufacturing processes for each component and, indeed, the manufacturing work centres which individual components must visit. PFA, according to Burbidge, is a progressive technique based on five sub-techniques, namely:

(1) Company Flow Analysis (CFA).
(2) Factory Flow Analysis (FFA).
(3) Group Analysis (GA).
(4) Line Analysis (LA).
(5) Tooling Analysis (TA).

CFA is used in multiplant companies to plan the simplest and most efficient inter-plant material flow system.

FFA is used to identify the sub-products within a factory around which product based departments can be organized.

GA is used to divide the individual departments into groups of machines which deal with unique *product* families.

LA seeks to organize the individual machines within a line to reflect the flow of *products* between those machines.

TA looks at the individual machines in a cell or line and seeks to plan tooling so that groups of parts can be made with similar tooling set-up.

The important issue from our point of view is that flow based manufacturing is an important goal for manufacturing systems designers to aim towards and it is certainly central to the whole JIT approach.

11.3.3 The relationship with suppliers in a JIT environment

As indicated earlier, the ideas of JIT are not restricted to the narrow confines of the manufacturing plant but also reach out to the factory's customers and back to the vendor companies who supply the factory with raw materials and purchased items. The approach is to build strong and enduring relationships with a limited number of suppliers, to provide those suppliers with the detailed knowledge they need to be cost effective to help them overcome problems they might encounter and to encourage them to apply their detailed knowledge of their own manufacturing processes constantly to improve the quality of the components they supply.

This involves taking a *long* term view of the buyer/supplier relationship and also involves commitment to building an enduring cooperative relationship with individual suppliers where information is readily shared and both organizations work to meet shared goals. JIT execution (i.e. the Kanban system), applied to purchasing, gives rise to frequent orders and frequent deliveries. The ideal of single unit continuous delivery (delivery lot size of one) is impractical but it can be approached by having as small a lot size as possible, delivered from the supplier, as frequently as possible. The physical distance of suppliers from the buyer's manufacturing plant plays an important role in determining the delivery lot size.

The closer the supplier is to the buyer's plant, the easier it is to make more frequent deliveries of smaller lots. Ideally, this may allow the supplier to initiate JIT production in his/her own plant and so link up with the buyer's JIT production system. In the case of suppliers at a distance from the buyer's plant, various techniques may be used to reduce what might otherwise be a high cost per unit load.

Consider for example the following situation. Four suppliers, A, B, C and D, respectively, supply components to buyer E. The four suppliers are located in relatively close proximity to each other but all are at a distance from the buyer E. If they must all deliver four times each day then the possibility exists for them to cooperate in such a way that deliveries are made to the buyer four times each day, with each supplier responsible for only one delivery run per day. Supplier A might make the first delivery picking up product from B, C and D *en route*. Supplier B could make the second delivery picking up the products of A, C and D, etc.

On one hand, the buyer places great demands on the supplier in terms of frequent or just in time deliveries of components. On the other hand, by providing the supplier with commitments for capacity over a long period and by ensuring that the supplier is aware of modifications to the company's master schedule as soon as is practicable, the company helps the supplier to meet the exacting demands of JIT. The ideal situation is where the supplier himself is able to focus a part of his/her plant to service each customer and the supplier starts to achieve a JIT environment in house.

The benefits from JIT purchasing include reduced purchased inventories, low rework, reduced inspection and shortened production lead time. These all combine to increase adaptability to demand and hence achieve just in time production.

11.4 Conclusion

This chapter has laid out the goals of JIT and also indicated what are considered to be the key elements in the JIT approach to manufacturing. These have been listed as:

- An intelligent match of market demand with product design in an era of greatly reduced product life cycles, with early consideration of manufacturing problems at the product design stage.
- The definition of product families based on a number of important manufacturing goals and the design of manufacturing systems to facilitate flow based production of these families, where possible.
- The establishment of relationships with vendors and suppliers to achieve just in time deliveries of raw materials and purchased components.

Chapter 12 will go on to look at some JIT manufacturing systems design techniques which are focused on the above goals and which may be used in the context of the JIT approach to manufacturing.

CHAPTER TWELVE

Manufacturing systems design and planning for JIT

12.1 Introduction

This chapter will focus on a set of manufacturing system design techniques that support the just in time approach. The design and planning of JIT manufacturing systems is applicable within the context of a just in time philosophy and is necessary to create the environment to allow Kanban, the shop floor control realization of JIT, to work as is illustrated in Figure 12.1.

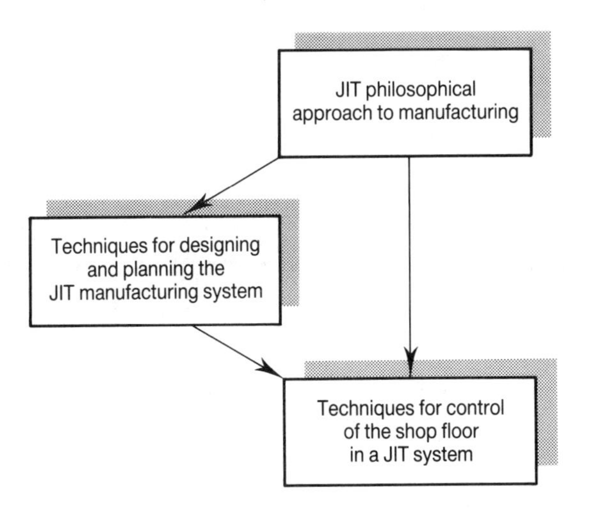

Figure 12.1 JIT approach.

The prime objective of this chapter is to review JIT techniques and to present the reader with an understanding of the role and importance of these techniques in JIT practice. These techniques relate primarily to manufacturing system design and production planning.

A primary focus of JIT is the reduction of the production lead time. Many advantages can be gained by reducing this lead time to a minimum. For example, a short lead time relative to competitors in the marketplace allows sales offices to quote shorter delivery times – an important competitive advantage in today's business environment. From the manufacturing planning perspective, short lead times reduce the manufacturing plant's dependence on forecasts and allow the plant to operate using a shorter planning horizon and, consequently, a more accurate master schedule. Reduced lead times also have important consequences for a plant's ability to respond to short term unexpected changes in the marketplace, as will be seen later.

There are many activities over the product life cycle and throughout the manufacturing enterprise which influence the product lead time, from product design right through to receipt of the completed order by the customer. Similarly, there are also approaches which improve manufacturing performance based on the reduced lead time. These approaches and techniques within JIT manufacturing can be grouped under five identifiable headings:

(1) Product design for ease of manufacture and assembly.
(2) Manufacturing planning techniques.
(3) Techniques to facilitate the use of simple, but refined, manufacturing control systems, namely Kanban.
(4) An approach to the use of manufacturing resources.
(5) Quality control and quality assurance procedures.

Figure 12.2 presents the techniques that support the just in time approach and facilitate the creation of the environment needed for JIT execution. Each of these will now be reviewed in turn.

12.2 Product design for ease of manufacture and assembly

Chapter 11 identified some key elements in the JIT approach, one of which is an intelligent match of the product design with the perceived market demand. It was noted that this is important in an era of constantly changing market demands when a manufacturer must offer a diversity of products within a given product range to the market.

As suggested in Chapter 11, what is required of the product design team, among other things, is that it interpret the wishes of the marketplace

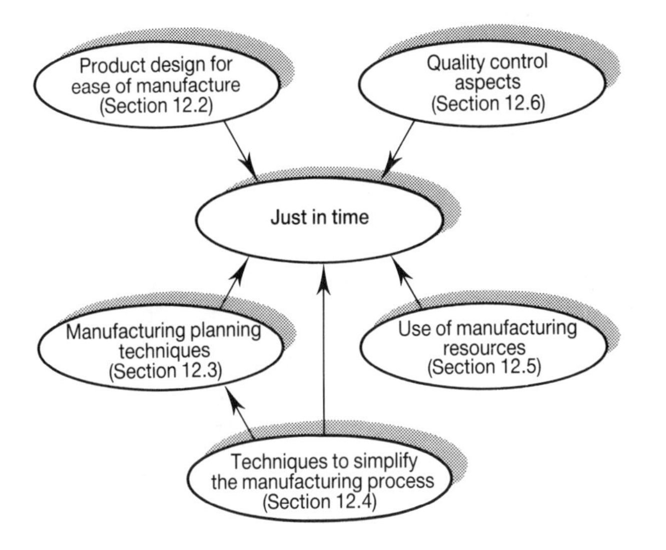

Figure 12.2 Manufacturing techniques which support JIT.

and, if possible, lead the market by introducing a product range that allows the production system to respond effectively to the market. This involves designing products that both anticipate the market requirement and include sufficient variety to meet consumers' expectations, while being manufactured at a price that the market is willing to pay. This can be achieved in many ways. One approach is to increase the variety of products offered without simultaneously increasing the required process variety, associated complexity and increased cost. Let us consider Figure 12.3.

At present, designers of manufacturing systems are required to move along a diagonal which is bounded by both economic and technological constraints as shown in Figure 12.3. The continuum of manufacturing mentioned in Chapter 1, which extends from mass production to jobbing shops, can be seen along this diagonal with mass production in the bottom right corner and jobbing shop production in the upper left corner. However, given the new *environment* of manufacturing discussed in Chapter 1, we might surmise that designers are attempting to move in the direction of low process variety and high product variety (i.e. the bottom left corner of Figure 12.3). This effort is bounded by technological constraints and the approach that seems to be prevalent in the West is often technologically driven, through the introduction of computer controlled and thus more flexible production facilities.

In our view, the just in time approach represents a more comprehensive attempt to move towards low process variety and high product variety. Not only concerned with technological improvements, JIT also utilizes such

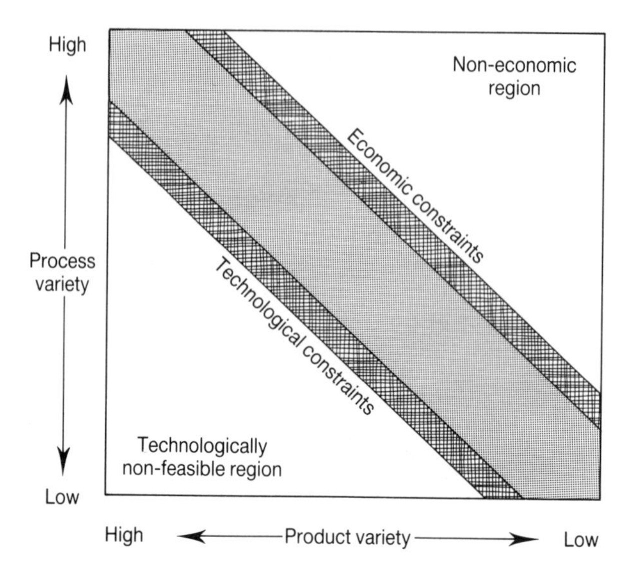

Figure 12.3 Tradeoff between product and process variety.

techniques as product design for manufacture and assembly, flexible equipment, a flexible workforce and superior production engineering practice in areas such as the design of jigs and fixtures, to achieve simple and, therefore, fast and inexpensive set-ups and changeovers between products. The concepts behind the use of flexible equipment and a flexible workforce will be brought out later in this chapter as will the JIT approach to set-up reduction. For the moment, the effect of short set-up times will be illustrated using the following example.

> A machine that manufactures two distinct products – A and B – can be considered, from the process viewpoint, to be producing one product if the set-up or changeover time between the two products is very small, in effect, approaching zero. To achieve this very desirable situation involves close collaboration between the product design, process engineering, and manufacturing people in the plant.

Traditional thinking about the interaction between product or process variety can be represented by Figure 12.4. From this perspective, a widening of the product range results in an increase in the process variety required to cope with the increased product options. If a manufacturing plant increases the options within its product range, this is normally expected to lead to increased process complexity because of increased process variety. Similarly, if the range of options in a product is reduced, this might be expected to lead to a reduction in the complexity of the production system. In fact, the

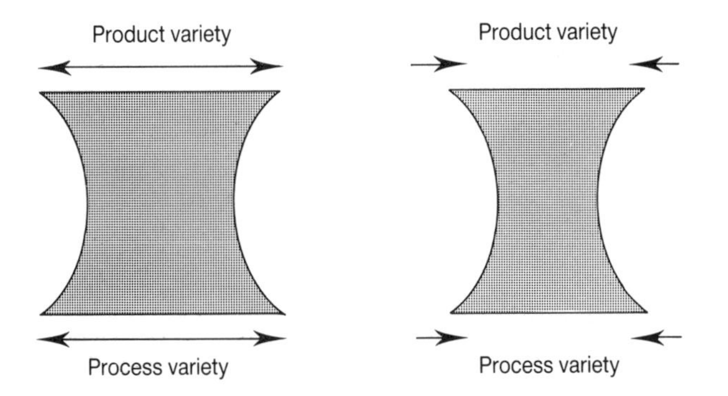

Figure 12.4 Product/process variety relationship (classical).

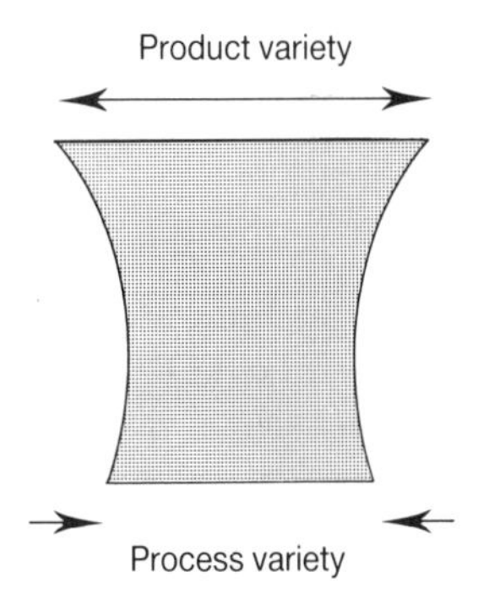

Figure 12.5 Product/process variety relationship under JIT.

classical distinction between mass production, batch production and jobbing shop production is based, at least partially, on this notion – mass production uses specialized equipment to manufacture a narrow range of products in high volumes efficiently. However, where the product range is large and each product is required in relatively small volumes, more general purpose equipment is required and the resulting process variety is high. In effect, economies of scale cannot be realized in batch based production systems. The impact of the introduction of computer based automation into plants has been to allow the manufacturer to deal with greater variety, but the basic underlying relationship has not changed. Increased product variety involves more process complexity and, therefore, increased cost.

However, the JIT approach tries, through intelligent product design and through consideration of process issues at the product design stage, to increase the variety of products within a manufacturing plant while maintaining, if not actually reducing, process variety (see Figure 12.5). (It should be

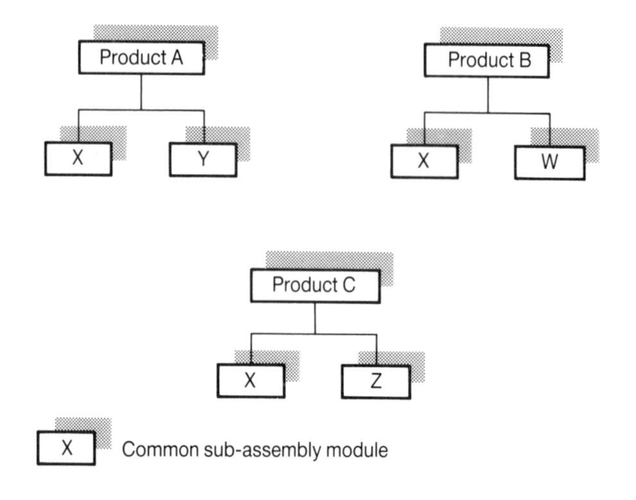

Figure 12.6 Common modules across BOMs.

pointed out that design for manufacture and design for assembly are not unique to JIT. What is unique about JIT, in our view, is the emphasis it places on these issues and the fact that process and product design seem to have equal *status* and to work together effectively.)

This can be achieved using techniques such as modular design, design for simplification and design for ease of manufacture and assembly. (See Treer (1979) for a more complete discussion on this important topic and its particular relevance when designing a product for automated manufacturing and assembly.)

12.2.1 Modular design

One of the consequences of good design is frequently a reduction in the number of components necessary to produce a given product and hence a reduction in the production lead time. Similarly, products may be designed in a modular fashion so that components and assemblies are common across a given product range and thus product variety is managed to good effect. With regard to the design process, it is possible to expand a basic model so as to increase the variety of products offered to the market. As illustrated in Figure 12.6, a common sub-assembly or module X is used across a number of products – A, B and C. The effect of this is to increase the requirement for a single module X rather than having three different modules, each with a relatively low total requirement. Thus the use of standardized components and sub-assemblies results in increased production volumes of fewer different components and, consequently, reduced inventory levels.

This design philosophy is also reflected in the bill of materials through attempts to keep the differences between products as *high as possible* in the product structure and thus minimize the consequences of variability for manufacturing.

12.2.2 Design for simplification

Design for simplification seeks to design products which are relatively simple to manufacture and assemble. New product designs should, as far as possible, include *off the shelf* items, standard items or components that are possible to make with a minimum of experimental tooling. Product features, such as part tolerances, surface finish requirements etc., should be determined while considering the consequences of unnecessary embellishment on the production process and thus on production costs. This approach can result in a major simplification of the manufacturing and assembly process.

12.2.3 Design for ease of automation

Design for ease of automation is concerned with the general concepts and design ideas which will, for example, in the case of assembled components, help to simplify the automatic parts feeding, orienting and assembling processes. In the case of assembled components it is important to design products to be assembled from the top down and to avoid forcing machines to assemble from the side or particularly from the bottom. The ideal assembly procedure can be performed on one face of the part, with straight vertical motions and keeping the number of faces to be worked on to a minimum.

In fact, it is the area of automated and, in particular, robot based assembly that the importance of the design for assembly approach can be most clearly seen. Until recently the application of robots in industry had been confined to relatively primitive tasks – machine loading and unloading, spot and arc welding, spray painting, etc. Relatively few applications in assembly have been realized. Researchers and manufacturing systems designers have adopted two main approaches: the development of sophisticated assembly robots and the redesign of products, components, etc. for robot based assembly. The first approach involves the development of *universal* grippers and *intelligent*, sensor-based robots with sufficient accuracy, speed and repeatability, and which are capable of being programmed in task oriented languages. This approach seeks to mimic the flexibility and power of the human arm and hand. The second approach seems to be the more successful in practice. Laszcz (1985) points out that '. . . a product designed in this manner reduces assembly to a series of pick-and-place operations, thereby requiring a less sophisticated robot. This results in manufacturing cost savings and increases the likelihood of financially justifying robotic assembly'.

12.3 Manufacturing planning techniques

An important purpose of JIT is clearly to reduce costs. This is achieved in many ways, the most notable being the elimination of all wastes, especially unnecessary inventories. For example, in sales, cost reduction is realized by supplying the market with first class products in the quantities required and at an affordable price. Stocks of finished goods are therefore minimized. To sell at a realistic price and in the quantities required, the production processes must be adaptable to changes in demand and be capable of getting the required products quickly through manufacturing and to the marketplace. Similarly, the warehouses must only stock materials in the quantities required. To help production to respond effectively to short term variations in market demand, Just in Time attempts to match the expected demand pattern to the capabilities of the manufacturing process and to organize the manufacturing system so that short term, relatively small, variations can be accommodated without major overhaul of the system. The technique used to help achieve this is known as production smoothing.

Through production smoothing, single lines can produce many product varieties, each day, in response to market demand. Production smoothing utilizes the short production lead times to *mould* market demand to match the capabilities of the production process. It involves two distinct phases as illustrated in Figure 12.7.

The first phase adapts to monthly market demand changes during the year, the second to daily demand changes within each month. The possibility of sudden large changes in market demand and seasonal changes is greatly reduced by detailed analysis of annual, or even longer term, projections and well thought out decisions on sales volumes, in so far as this is possible.

12.3.1 Monthly adaptation

Monthly adaptation is achieved though a monthly production planning process, i.e. the preparation of a Master Production Schedule (MPS), similar to the MRP process documented in Chapters 5 and 6. This MPS gives the averaged daily production level of each process and is typically based on an aggregate three month and a monthly demand forecast. The precise planning horizon depends very much on the industry in question – in the automotive industry where JIT originated, three months is typical. Thus the product mix and related product quantities are *suggested* two months in advance and a detailed plan is *fixed* one month in advance of the present month. This information is also transmitted to suppliers in order to make their task of providing raw materials as required somewhat easier. Daily schedules are then determined from the master production schedule.

In fact, the concept of production smoothing extends along two dimensions – firstly, by spreading the production of products evenly over each day within a month and, secondly, by spreading the quantities of each

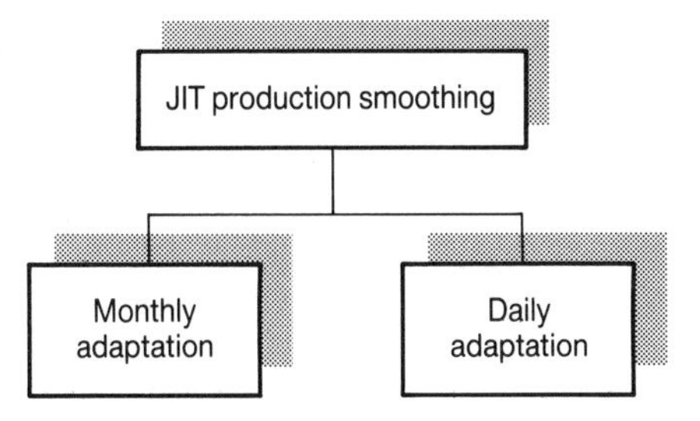

Figure 12.7 Components of production smoothing.

product evenly over each day within a month. Both of these are typically incorporated into the daily schedule as in the following example.

Consider a production line in, say Gizmo-Stools Inc., which produces six different stools – Stool A, Stool B, Stool C, Stool D, Stool E and Stool F. These stools have different characteristics – some are three legged stools, others are four legged stools, some have round seats while others have square seats, etc. Let us also assume that the master production schedule calls for 4800 units to be produced in a month containing 20 working days. Then, by averaging the production of all stools over each day, 240 units must be produced per day. If the 4800 stools breaks down into the product quantities in Table 12.1(a) then, in the extreme case of traditional batch production, the floor schedule would produce 1200 of Stool A, followed by 400 of Stool B, 1600 of Stool C, 400 of Stool D, 600 of Stool E and finally 600 of Stool F. However, by averaging the output of each product over all days within the month and assuming that there are 20 working days in the month the daily production schedule illustrated in Table 12.1(b) is calculated.

Therefore, the required 240 units must be produced within a shift (i.e. 8 hours or 480 minutes). Simple mathematics tells us that one unit must be produced every two minutes. This may be done in a batch of 60 of Stool A followed by a batch of 20 of Stool B, and so on. However, by carrying the second concept further and spreading the production of all products evenly within each day, we can develop a schedule for a small duration, e.g. 48 minutes, as in Table 12.2. This schedule is continuously repeated until the daily schedule is met.

12.3.2 Mixed model production

This concept of manufacturing and assembling a range of products *simultaneously* is known as **mixed model** production and it is widely used within what are termed repetitive manufacturing systems. Repetitive manufacturing

Table 12.1 Monthly and daily product quantities.

	(a)	(b)
Varieties	**Monthly demand**	**Daily average output**
Stool A	1200	60
Stool B	400	20
Stool C	1600	80
Stool D	400	20
Stool E	600	30
Stool F	600	30
Demand	4800 per month	240 per day

Table 12.2 Production schedule.

Varieties	Number of units
Stool A	6
Stool B	2
Stool C	8
Stool D	2
Stool E	3
Stool F	3
	24 every 48 minutes

systems will be examined in more detail in Chapter 13. Mixed model production should be differentiated from **multi-model** production where a variety of models are produced but not simultaneously. Mixed model production is clearly not feasible unless the set-up times for individual models are extremely small so that there is no effective changeover in going from, say, Stool A to Stool B. This, in turn, can only be achieved if the designs of the products in question are such that they minimize *process variety* (see Section 12.2).

The benefits of mixed model assembly are potentially very great, particularly in an environment where customers expect rapid turn around on orders and where the ability to respond at short notice is critical. Mixed model production offers a very high level of flexibility compared to traditional production methods.

Consider the following situation. Let us assume that Table 12.1 reflects the expectation of requirements for the six products – Stool A, Stool B, etc. at the start of the month. Based on this we have developed a

Table 12.3 Half-monthly and daily product quantities.

Varieties	Monthly demand	Daily average output
Stool A	600	60
Stool B	200	20
Stool C	800	80
Stool D	200	20
Stool E	300	30
Stool F	300	30
Demand	2400 per rest of month	240 per day

Table 12.4 Revised half-monthly product quantities.

Varieties	Monthly demand	Daily average output
Stool A	600	60
Stool B	600	60
Stool C	400	40
Stool D	200	20
Stool E	300	30
Stool F	300	30
Demand	2400 per rest of month	240 per day

basic production cycle as shown in Table 12.2. Now, further assume that in the middle of the month a major customer changes its order – for example, an important customer decides to change the order from 400 of Stool C to 400 of Stool B. How can this be accommodated? The expected requirement for the second half of the month has now been changed from that defined by Table 12.3 to that of Table 12.4.

Thus to meet this new situation, all that is required is that the planner modify the production cycle in line with the new mix within the daily output. In effect, all that is required is that the cycle represented by Table 12.2 is modified to that contained in Table 12.5.

This is clearly a trivial example but the point that it seeks to illustrate is nevertheless valid. Mixed model production results in a flexible production system and one which is very responsive to sudden market changes. What would the situation have been if the system described above had been operated in the traditional manner where, admittedly in an extreme case, all of Stool A might have been manufactured first followed by all of Stool B, etc.?

Table 12.5 Revised production schedule.

Varieties	Number of units
Stool A	6
Stool B	6
Stool C	4
Stool D	2
Stool E	3
Stool F	3
	—
	24 every 48 minutes

12.3.3 Daily adaptation

After the development of a monthly production plan, the next step in the smoothing of production is the breakdown of this schedule into the sequence of production for each day. This sequence specifies the assembly order of the units to be produced. The sequence is arranged so that when the cycle time expires, one group of units has been produced. At every work centre no new units are introduced until one is completed. This sequence schedule is *only* transmitted to the starting point of final assembly. Kanban (see Chapter 13) functions so that production instructions are simply and clearly transferred to all other assembly and manufacturing processes.

Referring back to Table 12.2, it can be seen that a mix of 24 units of the four stools must be produced within 48 minute intervals. The sequence of production for these 24 units might be be as follows:

AAAAAA BB CCCCCCCC DD EEE FFF.

Or the sequence could be more varied, such as:

ACAECFBCACDAEFCBCADCFEAC.

Attaining the *optimal* sequence is difficult. Heuristic procedures have been developed (Monden (1983) and Wild (1984)) which produce a sequence that aims to achieve two plantwide goals:

(1) An even load at each stage of the manufacturing process.
(2) A constant depletion rate for each component.

In greatly simplified terms, the objective of the heuristic is to minimize the variations in consumed quantities of each component at final assembly and at all of the work centres. By this smoothing of production, large fluctuations in demand and the amplification of these fluctuations back through the production system are prevented.

Each day's schedule should resemble the previous day's schedule as closely as possible. Hence, uncertainties are eliminated and the need for dynamic scheduling and safety stocks is minimized.

The daily adaptation to the actual demand for varieties of a product during a month is the ideal of JIT production, which in turn requires the daily smoothed withdrawals of each part from the sub-assembly lines right back to the suppliers. Minor variations in demand are generally overcome by the Kanban system by increasing or decreasing the number of cards. This will be fully discussed in Chapter 13.

One might argue that the ideal of production smoothing is, in fact, very difficult to achieve in practice for many industries and individual companies. Nevertheless, it is clear that for all of manufacturing industry there are lessons to be learned from this approach. The key lesson is the fundamental importance of a firm master schedule from the point of view of control of the production system and the advantages, in terms of flexibility, to be gained from moving towards mixed model production and assembly. It was seen during the discussion on the master schedule (Chapter 7) that the longest cumulative lead time determines the length of the planning horizon. Section 12.4 will now go on to see how JIT seeks to reduce throughput times or lead times and hence reduce the planning horizon, which in turn lessens the dependence on forecasting in the master schedule development.

12.4 Techniques to simplify the manufacturing process and reduce lead times

As we discussed in Chapter 10, the lead time or throughput time for a batch through the shop floor is typically much greater than the actual processing time for the batch in question. It is not unusual in conventional batch manufacturing systems for the actual processing (including set-up time) to represent less than 5% of the total throughput time. Furthermore, of that 5%, only 30% may be spent in value adding operations. The throughput time or lead time for a product is composed of four major components – the actual process time including inspection time, the set-up time, the transport time and the queuing time as illustrated in Figure 12.8. In a typical batch manufacturing shop, this latter component is frequently the largest, often representing in excess of 80% of total throughput time.

Heard and Plossl (1984) describe the elements of lead time and their inter-relationships. They recommend techniques to reduce these elements. JIT has long understood this and attempts to reduce each of these individual elements as much as possible.

The single largest element of throughput time in traditional batch manufacturing systems is the queuing and transport time between operations.

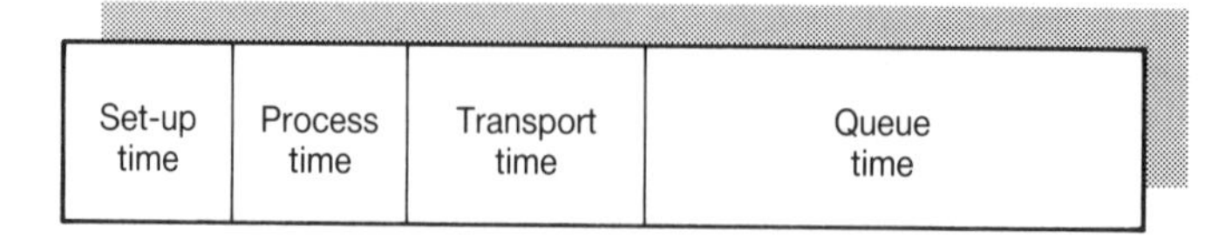

Figure 12.8 Breakdown of the lead time.

As discussed in Chapter 11, JIT encourages product based plant layouts which greatly reduce throughput times for individual batches by reducing queuing time. At a plant level, the product based layout reduces throughput time by facilitating easy flow of batches between operations and work centres. At a line or work centre level, JIT reduces throughput time by using what are termed **U-shaped** layouts. In our view, the effort to reduce throughput time must be seen in the context of the product based layout at the macro-level and the U-shaped layout at the work centre level.

How JIT addresses each of the elements of throughput time will be discussed later. Section 12.4.2 will focus on the JIT approach to the reduction of queuing time, Section 12.4.3 will consider transport time, Section 12.4.4 will consider set-up time reduction and, finally, Section 12.4.5 will look briefly at processing time. Before going on to this however, the U-shaped layout will be considered in some detail.

12.4.1 Layout of the production processs

The major objectives in designing the production layout at the work centre level are similar to the plant level objectives and can be listed as follows:

- To provide flexibility in the number of operators assigned to individual work centres in order to be able to adapt to small changes in market demand and consequently in the schedule.
- To utilize the skills of the **multifunction** operators.
- To facilitate movement towards **single unit production and transport** between work centres.
- To allow for the re-evaluation and revision of the standard operations.

To meet these objectives, the U-shaped product based layout was developed as illustrated in Figure 12.9 (adapted from Monden (1983)). This layout allows assignment of a multifunction operator to more than one machine due to the close proximity of the machines.

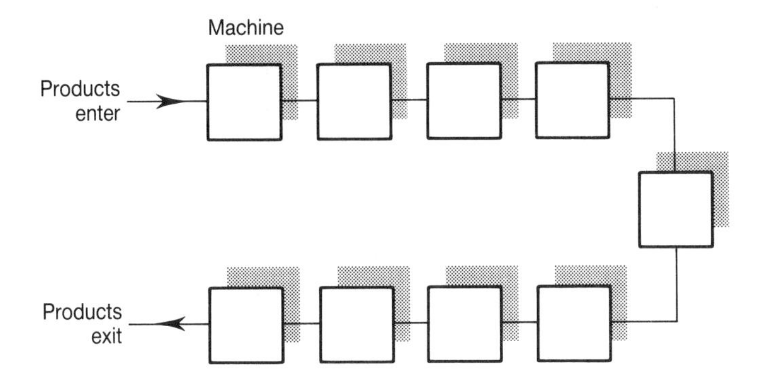

Figure 12.9 U-shaped work cell layout.

Using this layout the range of jobs that each operator performs may be increased or decreased, allowing flexibility to increase or decrease the number of operators. It allows *unit production and transport*, since machines are close together and may be connected with chutes or conveyers. Synchronization is achieved since one unit entering the layout means one unit leaving the layout and going on to the next work centre.

The use of the U-shaped layout with its requirement for multiskilled operators clearly implies an increased need for operator training, as well as for very well defined and documented manufacturing instructions for operators. It is implicit in the JIT approach to manufacturing that no effort or expense be spared in training; where necessary, retraining operators in highly tested and refined work practices is considered.

12.4.2 Reduction of the queue time

Within the context of product based and U-shaped layouts, various techniques are used to reduce queue time. Figure 12.10 lists some of these techniques which will now be considered briefly.

Small production and transport lots

In just in time manufacturing, one unit is produced within every cycle time and at the end of each cycle time a unit from each process in the line is simultaneously sent to the next process. This is already prevalent in the assembly line systems of virtually all companies engaged in mass production. However, processes supplying parts to the assembly lines are usually based on lot production. Just in time, however, seeks to extend the concept of *unit*

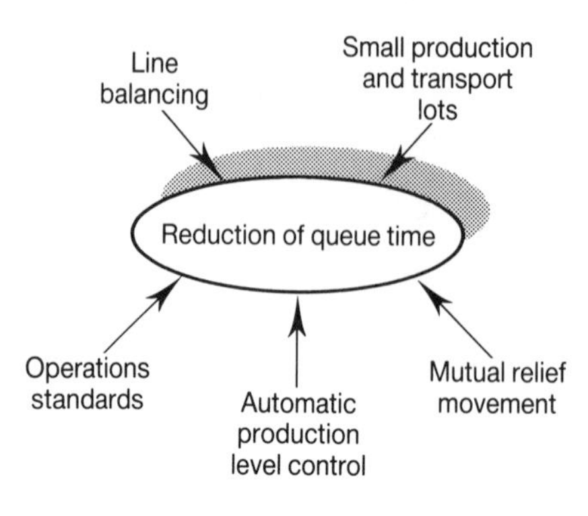

Figure 12.10 Methods of reducing queue time.

production and transport to processes such as machining, welding, pressing, etc. which feed the final assembly lines. Therefore, as in an assembly line, operations must start and end at each process at exactly the same time. This is often called **synchronization**, i.e. continuous flow production.

In our discussion of the lead time and the MRP approach to production and inventory management, it was pointed out that MRP seems to consider only one batch or lot size, namely the production lot. The advantage of distinguishing between the production and conveyance lots was alluded to. For example, in situations where large production lots are necessary because of large and irreducible set-up times, smaller conveyance lots can be used to reduce overall throughput times. This notion is important to JIT manufacturing and worth restating.

Consider three operations with a cycle time of one minute each. One unit would take three minutes to go through all operations. If batch production is employed and the process lot is 200 then total throughput time is (200 + 200 + 200) minutes or 10 hours. In this case the transfer lot is equal to the process lot as in Figure 12.11. In simple terms, a single lot size is used and lots are not normally split to facilitate early dispatch of partial lots to subsequent operations.

However, if the transfer lot (conveyance or transport lot) is less than the process lot, say, in the ultimate lot size one, then the total throughput time is greatly reduced as illustrated in Figure 12.12.

In fact, the total throughput time is 3 hours and 22 minutes. The total processing time is, of course, unaffected. In effect, the queue time has been greatly reduced.

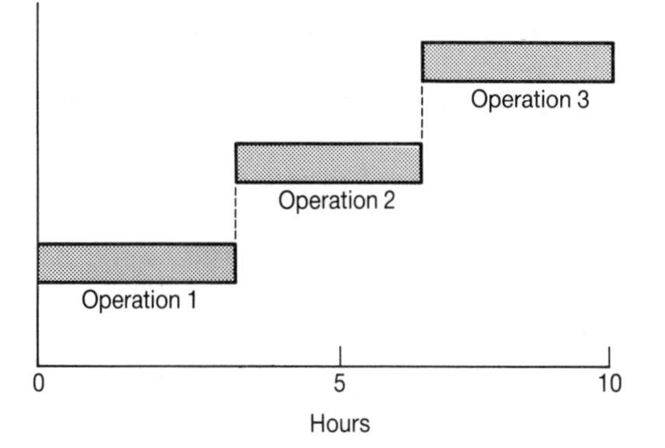

Figure 12.11 Transfer lot equals process lot.

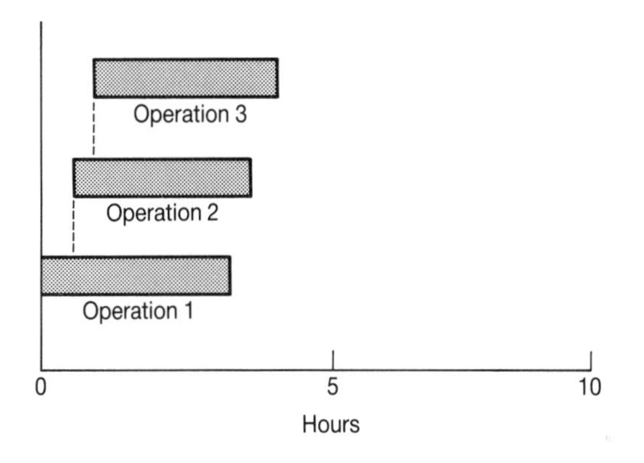

Figure 12.12 Transfer lot not equal to process lot.

JIT, in separating production lots from transport lots in situations where production lots are large, is seeking to move away from batch based production systems and towards flow based systems.

Line balancing

Line balancing seeks to reduce the waiting time caused by unbalanced production times between individual work centres and ensures production is the same at all processes, both in quantities and timing. Variances in operators' skills and capabilities are minimized, generating in advance well thought out and documented standard operations and by ensuring that all operators are trained in these *optimum* operation methods. What variances remain are smoothed through **mutual relief** (which will be discussed later). Line balancing is also promoted by the automatic control of production levels and *unit production and transport*. Synchronization also helps to balance the production timing between processes and facilitates line balancing.

It is interesting to compare briefly the notion of line balancing as seen from within a JIT perspective and the so-called *line balancing problem* well known to generations of students of industrial engineering. If we look at a typical Western textbook covering operations management and production systems design, we will, almost always, find a section on line balancing techniques. (See, for example, Wild (1984) and Groover (1980)). Line balancing is presented as a problem to be solved using algorithmic or heuristic procedures which seek to minimize what is termed the *balance loss* or some similar measure. The procedure is to take assumed elemental operations and operation precedence constraints and allocate the operations to assembly stations in order to divide the total work content of the job as evenly as possible between the assembly stations. The interesting point is that the emphasis is placed on allocating predefined operations to stations. The JIT approach places greater emphasis on the design of the operations and on ensuring that individual operators are skilled in carrying them out. Only then is the *analytical* approach to allocating the operations between stations brought to bear on the problem.

Automatic production level controls

In a particular work centre, a situation may exist where two machines operate on the same product. If the machine performing the first operation has a greater capacity than the second machine, it would traditionally build up a safety stock before the second machine. However, the JIT approach would couple both machines, and the first machine would only produce when the number of parts between the two machines was below a predefined minimum. It continues producing until the queue between the machines has reached a predefined maximum. This reduces the safety stock between machines and also reduces the queue times of products. This concept and its relation to Kanban will be discussed in Chapter 13.

Operations standards

Standardizing the operations to be completed at work centres attempts to attain three goals:

(1) minimum work in progress,
(2) line balancing through synchronization within the cycle time,
(3) high productivity.

Creating operations standards is a three stage procedure – determination of the cycle time, specification of operations for each operator and, finally, specification of a minimum quantity of WIP to allow smooth production.

Table 12.6 Determination of cycle time.

$$\text{Cycle time} = \frac{\text{Available daily production time}}{\text{Required daily output}}$$

The cycle time is determined, as illustrated in Table 12.6, with no allowance for defective units, down time or idle time in the available daily production time.

For each component/subassembly at every work centre, the completion time per unit is determined including manual and machine elements. By taking into account the number of components required for each finished product, the cycle time for the product and the completion times for each component, a list of operations for each operator is generated. This list of operations specifies the number and order of operations an operator must perform within the cycle time. This ensures production of the correct number of components/sub-assemblies to allow the production of one finished product within each cycle time.

Finally, the minimum quantity of WIP necessary to ensure production without material shortages is specified. This incorporates the minimum material between, and on, machines, that is required for continuous production.

Once the three phases (the cycle time, the order of operations and the standard WIP levels) are completed, they are combined to give a standard operations sheet which is then displayed where each operator can see it. With each new master planning schedule, gross estimates are presented to all processes of the demands likely to be made on them. At this point re-evaluation of operations standards, through reassignment of tasks to operators, for example, may result in a reassignment of the workforce to meet the projected requirement.

In this discussion on the U-shaped layout, the need for good manufacturing documentation and for constant training and retraining of operators in good work practices has been emphasized. This approach to generating operations standards clearly facilitates this documentation and training. In turn, well trained operators ensure that operations standards are adhered to.

Mutual relief movement

As will be seen in Section 12.5.1, operators in a JIT environment tend to be very versatile and are trained to operate many different machines and carry out many operations within their particular work centre, e.g. one operator may be able to operate a drilling machine, a lathe and a milling machine. As the plant and equipment layout are product oriented, this means that advantage can be taken of the multiskilled operator. The multifunction

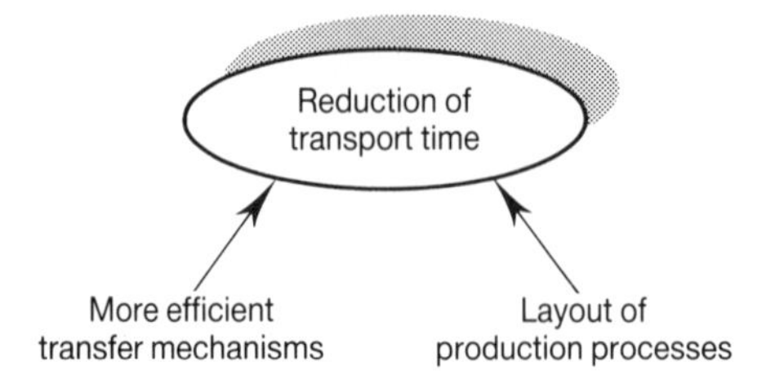

Figure 12.13 Reduction of the transport time.

operator also helps to reduce, if not eliminate, inventories between processes since when he/she unloads a part from one machine he/she may load it directly onto the next machine.

Operators regularly help each other on the shop floor. The ANDON (see Section 12.6) allows an operator to call for help if he/she is in difficulty. Since the work centres are close together and the operators are multifunctional, this mutual support is feasible. Because an operator can go to the aid of a colleague who is temporarily overloaded, the queues in front of work centres can be reduced and the effect of what would otherwise constitute a bottleneck in the system is alleviated, thus reducing the overall queuing time.

12.4.3 Reduction of the transport time

Figure 12.13 depicts two techniques that help to reduce the transport time in the just in time approach, namely the layout of the production processes and faster methods of transport between production processes. We have already discussed the product oriented system of plant layout and the U-shaped layout of equipment which tend to minimize transport needs between individual operations on a component or assembly.

It should, of course, be remembered that approaching of *unit production and transport* will most likely increase the transport frequency, i.e. the number of transports of partially completed units between operations. To overcome this difficulty, quick transport methods must be adopted, together with improved plant and machine layout. Belt conveyors, chutes and forklifts may be used. Generally, the close proximity of the subsequent process, as determined by plant layout, results in a minimum transport time between operations.

12.4.4 Reduction of set-up time

A major barrier to the reduction of the processing time and the ability to smooth production is the problem of large set-up times. The EOQ model has already been discussed (see Chapter 9) and it is graphically represented in Figure 12.14.

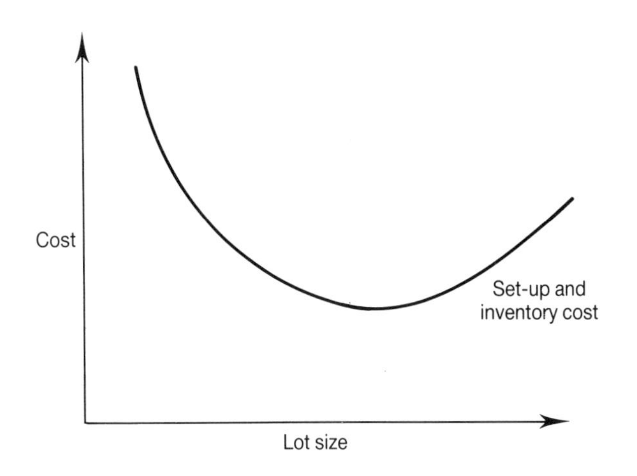

Figure 12.14 Economic order quantity model.

As noted earlier, the economic order quantity model seeks to determine a lot size which marks an optimum tradeoff between set-up and carrying costs in the case of manufactured items. EOQ calculations result in large lot sizes when set-up times and costs are high. However, large lot sizes and the resulting buffer stocks are incompatible with the JIT approach and hence the concentration of effort on reduction of set-up times. This makes a smaller lot size feasible and is a step on the road to *unit production and transport*.

This JIT approach contrasts strongly with EOQ thinking and cultivates the idea that machine set-up time is a major source of waste that can, and should, be reduced. The influence of set-up time reduction is illustrated by the fact that if set-up time is reduced to $1/N$ of its initial value, the process lot can be reduced to $1/N$ of its initial value, without incurring extra costs.

The throughput time is therefore reduced, work in progress inventory is reduced and the ability to produce many different varieties is enhanced. This makes for better response to demand. Likewise, the ratio of machine utilization to its full capacity will be increased without producing unnecessary inventory. In this way, productivity is enhanced. The techniques and concepts involved in reducing set-up are now briefly examined.

In order to shorten the set-up time, JIT offers four major approaches (Monden 1983):

(1) Separate the internal set-up from the external set-up. Internal set-up refers to that element of the set-up process which requires that the machine be inoperative in order to undertake it.

(2) Convert as much as possible of the internal set-up to the external set-up. This is probably the most important practical approach for the reduction

of set-up in practice and helps to achieve the goal of **single set-up**. Single set-up means that the set-up time can be expressed in terms of a single digit number of minutes (i.e. less than 10 minutes).

(3) Eliminate the adjustment process within set-up. Adjustment typically accounts for a large percentage of the internal set-up time. The reduction of adjustment time is, therefore, very important in terms of reducing the total set-up time.

(4) Abolish the set-up where feasible.

In the JIT approach to reducing set-up, the first step is to carry out a detailed study on existing practices. Internal and external set-up invariably overlap and need to be rigorously separated. A written specification outlining the procedures involved in the set-up and giving any necessary information is drawn up. By converting as much as possible of the set-up time to external set-up, which can be carried out offline, a significant improvement in internal or *at the machine* set-up time can be achieved.

The reader interested in a more detailed discussion on this topic of reducing set-up time is referred to Shingo (1985). He discusses the SMED system. SMED is an acronym for *Single Minute Exchange of Dies*, which connotes a group of techniques used to facilitate set-up operations of ten minutes and under.

12.4.5 Processing time

The JIT approach to processing time is relatively straightforward. It sees *processing time* as the only time during a product's passage through the production system that real value is actually being added to it. Transport time, queue time and set-up time are seen as non-value adding and to be reduced to the absolute minimum where they cannot be simply eliminated. Given that the processing time represents value added, JIT takes care to ensure that this time is used to the best advantage and to produce high quality product efficiently. Thus, as seen earlier in the discussion on the *U-shaped layout* and *operations standards*, great care is taken to ensure that the best manufacturing methods are refined to the highest degree, documented and communicated to the operators concerned through training sessions, practice sessions, etc.

In fact, it is in the commitment to the best possible manufacturing methods practised by a skilled and trained workforce that the pursuit of excellence in the JIT approach to manufacturing can be most clearly seen. So often in the conventional approach to manufacturing operations, manufacturing analysts and engineers forget the importance of good practice at the sharp end of manufacturing, namely the shop floor, and come to accept unnecessary deviation in operator performance as natural.

12.4.6 Concluding remarks on manufacturing process simplification

All the above ideas and techniques combine to reduce the overall throughput time and lead time for products. This results in order oriented production, a shorter master schedule planning horizon (remember that the master schedule planning horizon depends on the item in the master schedule with the longest cumulative lead time) and, consequently, a greatly reduced dependence on forecasts. Production smoothing is simplified if the production lead time is short. The ability to adapt to short term changes in market demand is enhanced and smooth withdrawals through the plant are easier to achieve.

12.5 The use of manufacturing resources

The JIT approach to the resources available in a manufacturing plant is interesting. The approach could be summarized in a single dictum – *do not confuse being busy with being productive*. This philosophy is particularly applied to the use of labour resource and is, as will be seen in Part IV, also fundamental in the OPT (Optimized Production Technique) approach to manufacturing.

The way which JIT seeks to use the major resources of labour and equipment efficiently and effectively will now be briefly reviewed.

12.5.1 Flexible labour

In JIT, those minor changes in demand which cannot be accommodated through the use of increased kanbans (see Chapter 13) are addressed through redeployment of the workforce. Adaptation to increased market demand can ultimately be met through the use of overtime.

However, JIT has a more subtle and effective approach to meeting relatively small short term demand changes. As mentioned in Chapter 11, the basic tenet of the just in time philosophy is the production of only those products that are required and at the precise time they are required. Under the principle of multifunction operators and multiprocess handling, one operator tends to a number of different machines simultaneously to meet this demand. Such a situation invariably results in the possibility of increasing output through introducing more operators into the system.

Therefore, if market demand increases beyond a level where increased kanban utilization is able to cope, temporary operators may be hired. Each operator may then be required to tend fewer machines thus taking up the equipment capacity slack. This approach presumes an economic and cultural environment where temporary operators of the required skill level are available and are willing to work in such a manner.

Adapting to *decreases* in demand is understandably more difficult, especially when one considers that many large Japanese companies offer lifetime employment. However, the major approaches are to decrease overtime, release temporary operators and increase the number of machines handled by one operator. This will cause an increase in the cycle time, thus reducing the number of units produced. Operators are encouraged to *idle* rather than produce unnecessary stock. They may be redeployed to practice set-ups, maintain and/or modify machines or to attend quality circle meetings.

The workforce is therefore flexible in two ways. It can be increased or decreased through temporary operators. It can also be relocated to different work centres. This latter flexibility demands a versatile, well trained multifunction operator as well as good design of machine layout.

The most important objective is to have a manufacturing system which is able to meet demand and to accommodate small, short term fluctuations in demand with the minimum level of labour. This does not imply the minimum number of machines. Companies operating JIT usually have some extra capacity in equipment, allowing for temporary operators when increased production is required.

It is interesting to note that few manufacturing facilities have used the JIT approach at the manufacturing system design stage, i.e. have been designed according to JIT principles. A manufacturing facility designed for the continuous production of as many components and/or assemblies and/or products as possible and not of the required amount of product at the right time as in JIT, frequently has some excess of capacity when examined from a JIT perspective.

12.5.2 Flexible equipment

Just in time requires production of different product varieties on the same assembly line each day (see the discussion on mixed model assembly in Section 12.3.1). This can involve a conflict between the market variety demanded by the customer and the production process available to service this market requirement since, in traditional manufacturing systems, it is normally desirable to reduce the variety of product going through the system. As has been seen, the JIT approach seeks to overcome these difficulties. Through consideration of process requirements at the product design stage, multifunction equipment is developed to help resolve this conflict by providing the production process with the ability to meet the variety demanded by the market. The specialized machines developed for mass production (as described in Chapter 2) are not suitable for repetitive manufacturing. By modifying these machines and adding minimum apparatus and tools they are transformed into multifunction machines capable of producing a product range that meets the marketplace demand. Such machines support just in time manufacturing and also facilitate production smoothing. Of course, the ultimate realization of this notion is the FMS (Flexible Manufacturing System) discussed earlier in Section 2.6.

This approach contrasts greatly with MRP manufacturing thinking where the capacity of resources are generally considered as fixed, at least in the short term. Hence in MRP, to meet demand variations, extra capacity is sought rather than attempting to modify the existing machines.

12.6 Quality control aspects of JIT

In more conventional production systems, work in progress inventories are often used to smooth out problems of defective products and/or machines. Batch production and the concept of Acceptable Quality Levels (AQL) could be seen to promote this attitude. This approach is criticized by the promoters of JIT thinking on the basis that it is treating the symptoms while not attempting to understand and resolve the underlying fundamental problems. In JIT manufacturing, the emphasis is on the notion of Total Quality Control (TQC) where the objective of eliminating all possible sources of defects from the manufacturing process, and thereby from the products of that process, is seen to be both reasonable and achievable.

It has been pointed out that JIT seeks *zero defects*. The zero defects approach involves a continuous commitment to totally eliminate all waste including, in this context, yield losses and rework due to product or process defects. The methods used to achieve zero defects are those of continuous steady improvement of the production process. Schneidermann (1986) offers an interesting analysis of the process of continuous improvement towards zero defects and suggests that it should be contrasted with an alternative improvement process – the innovation process. On the one hand, the continuous improvement route involves groups seeking small steps forward on a broad range of issues, using the available know-how within the group. The innovation process, on the other hand, seeks to achieve *great leaps forward* in narrowly defined areas through the use of science and technology by well qualified individuals. Here again there is evidence of the JIT approach being a systems approach with clear emphasis on involvement by all of those directly concerned.

Inspection is carried out to *prevent* defects rather than simply *detect* them. Machines, in so far as possible, are designed with an in-built capability to check all of the parts they produce as they are produced. The term **autonomation** was coined to describe this condition. This can be considered as one step on the road to total systems automation (i.e. a machine finds a problem, finds a solution, implements it itself and carries on).

Autonomation suggests automatic control of defects. It implies the incorporation of two new pieces of functionality into a machine:

(1) a mechanism to detect abnormalities or defects, and
(2) a mechanism to stop the machine or line when defects or abnormalities occur.

When a defect occurs the machine stops, forcing immediate attention to the problem. An investigation into the cause(s) of the problem is initiated and corrective action is taken to prevent the problem from recurring. Since, through autonomation, machines stop when they have produced enough parts and also only *good* parts, excess inventory is eliminated thus making JIT production possible.

The concept of autonomation is not limited to machine processes and *autonomous* checks for abnormal or faulty product can be extended to manual processes, such as an assembly line, using the following approach. Each assembly line is equipped with a call light and an ANDON board. The call light has different colours signifying the different types of assistance and support which might be required. It is located where anybody who might be called on to support the process (e.g. supervisor, maintenance, nearby operators, etc.) can easily see it.

The ANDON is a board which shows which operator on the line, if any, is having difficulties. Each operator has a switch which enables him/her to stop the line in case of breakdown, delay or problems with defective product. In many cases there are different colours to indicate the condition of the station on the assembly line which is having problems. The following are some colour signals that might be used and their respective meaning:

- red – machine trouble,
- white – end of a production run,
- green – no work due to shortage of materials,
- blue – defective unit,
- yellow – set-up required.

When an ANDON lights up, nearby operators quickly move to assist and solve the problem, and the supervisor takes the necessary steps to prevent it recurring. The ANDON also helps to ensure that completed products exiting the assembly line do not need rework, i.e. that they are right first time. Individual operators have *line stop* authority to ensure compliance with standards. Hence, the overall quality level is increased since each individual operator is encouraged to accept responsibility for the quality of the parts which he/she is involved with.

There are numerous other factors which assist in attaining extremely high quality levels. Small lot sizes, for example, will highlight quality problems very quickly as individual items are rapidly passed to the next process and any defects are quickly detected. Similarly, a good approach to *housekeeping* is encouraged and is considered important since a clean, well maintained working area leads to better working practices, better productivity and better personnel safety. Preventative maintenance is an important concept of the JIT approach. Using the *check-list technique*, machines are checked on a regular basis and repairs/replacements are scheduled to take place outside working time. This, in turn, helps to increase machine availability.

As a result of autonomation only 100% good units are produced. Hence, the need for rework and buffer or *insurance* stocks is eliminated. This lends itself to adaptability to demand and JIT production.

12.7 Conclusion

This chapter has presented an overview of the ideas and techniques used to create the manufacturing environment within which JIT execution can be achieved. Clearly, within the JIT approach to manufacturing, great emphasis is placed and enormous planning and engineering effort is expended to ensure that the manufacturing environment is such that excellence can be achieved. The JIT approach to manufacturing systems design and operation has been considered in terms of a number of specific issues:

- product design for ease of manufacture and assembly,
- manufacturing planning techniques,
- techniques to facilitate the use of simple manufacturing control systems,
- an approach to the use of manufacturing resources,
- quality control and quality assurance procedures.

The JIT approach recognizes the importance of process and manufacturing system design. Perhaps, in the West, this work has been neglected and the organization of our manufacturing facilities reflects this neglect. The following chapter will describe the Kanban card system.

CHAPTER THIRTEEN

The Kanban system

13.1 Introduction

This chapter focuses attention on the shop floor implementation of just in time. The techniques used at this level have been well documented in the last few years (APICS (1984), Monden (1981), Schonberger (1982) and Monden (1983)) as interest in Japanese manufacturing techniques has increased. The system that executes JIT delivery on the shop floor level is known as Kanban and the cards that are used in this system are called kanban cards. Therefore, to distinguish between the system and the cards we will use 'Kanban' for the system and 'kanban' for the cards. The discussion will concentrate on JIT execution on the shop floor and will not cover JIT execution from outside suppliers in any detail.

Before analyzing the operation of the kanban cards, the difference between **push** and **pull** systems of production control will be discussed, together with the notion of a repetitive manufacturing system, within which the kanbans can be used to greatest effect.

13.2 Kanban

Kanban was developed at the Toyota car plants in Japan as a program to smooth the flow of products throughout the production process. Its aim is to improve system productivity and to secure *operator involvement* and participation in achieving this high productivity by providing a *highly visible* means to observe the flow of products through the production system and the build up of inventory levels within the system. Later it was further developed as a means of production activity control to achieve the goals of JIT and to manage the operation of just in time production. Kanban also serves as an information system to monitor and help control the production quantities at every stage of the manufacturing and assembly process.

Kanban is seen as a *pull* system, as distinct from the production activity control systems in MRP which are regarded as *push* systems. Before discussing Kanban in detail, the differences between pull and push systems of production control will briefly be explored.

13.2.1 Pull system of manufacturing management

As stated above, the operation of MRP at the shop floor level is best described in terms of a *push* system, as distinct from Kanban which is considered to be a *pull* system. Considering the difference between the two approaches may help bring out some of the essential characteristics of Kanban.

Both systems are driven by a master schedule which defines the requirement for individual products, i.e. top level items in the bill of materials. This master schedule, in turn, is broken down into a detailed plan for items to be manufactured, assembled and purchased.

A push system operates as indicated in Figure 13.1 (Menga 1987). Let us consider, for a moment, a component which must be processed through a series of work centres, namely M through to 1, where work centre M processes the item first, followed by work centre M-1, etc. In a *push* system, work centre M is given the due date (DD) for the item and the item is released for production at the release time (RT). The completion time for work centre M becomes the release time for work centre M-1, etc. Thus, a product is pushed through the production system starting at the release of raw material to the first processing work centre and leading onto completion at the final work centre. MRP is considered to be the classical example of a push system.

A *pull* system, on the other hand, looks at the manufacturing process from the other end, i.e. from the perspective of the finished item. The production controller works on the basis that his/her orders represent firm customer requirements. The time horizon is understandably short. The orders are broken down from the highest level and the controller checks whether sufficient component parts are available to produce the finished product. If the components are available, the product is produced. However, if they are not, components are *pulled* from the preceding work centre. A similar procedure is followed right back through each production stage and extending all the way back to include outside vendors. Such a system places great demands on the production system and vendors. These demands can be

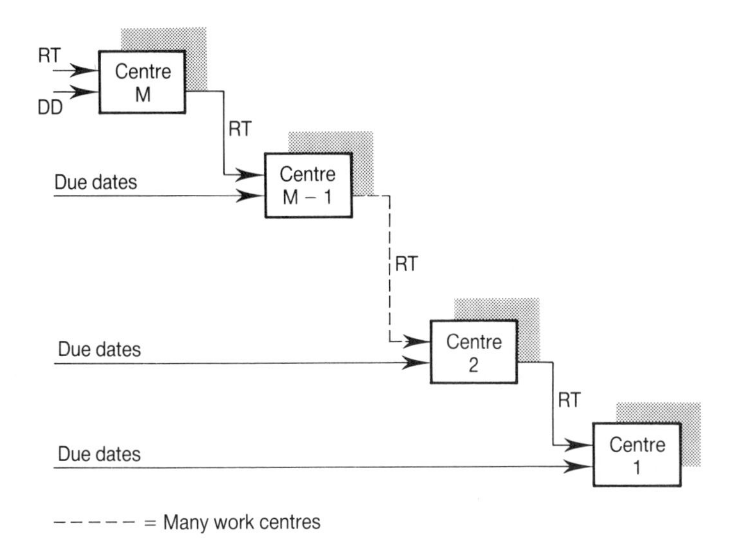

Figure 13.1 MRP 'push' strategy.

met either by having all component parts in inventory or, alternately, having the capability to respond and make them available in a very short time (i.e. short lead time). A pull system is illustrated in Figure 13.2.

It should be noted, and this will become more obvious as the operation of Kanban is reviewed in more detail, that this type of control mechanism is only applicable in plants involved in what is termed by many manufacturing systems analysts as **repetitive manufacturing**. Let us quickly define repetitive manufacturing and then go on to look at how the kanban cards are used.

13.3 Repetitive manufacturing

> Repetitive manufacturing is 'the fabrication, machining, assembly and testing of discrete, standard units produced in volume, or of products assembled in volume from standard options . . . [it] is characterized by long runs or flows of parts. The ideal is a direct transfer of parts from one work centre to another' (Hall 1983).

Referring back to the discussion on the various categories of discrete parts manufacturing system in Chapter 1, repetitive manufacturing can be positioned in the modified version of Figure 1.5 shown in Figure 13.3.

One could argue that the end result of rigorously applying the JIT approach and of using JIT manufacturing techniques as described in

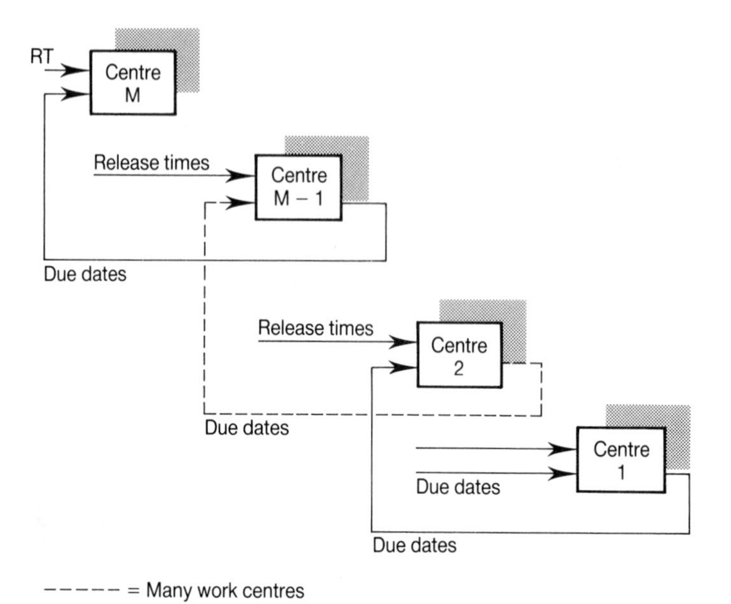

Figure 13.2 JIT 'pull' strategy.

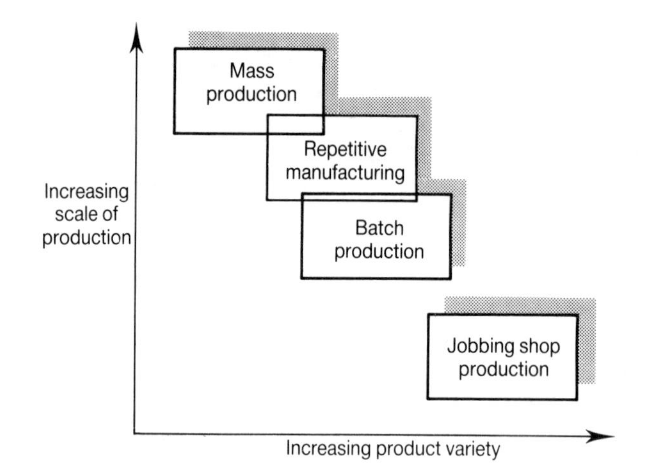

Figure 13.3 Classification of discrete production.

Chapters 11 and 12 is to move a manufacturing system away from jobbing shop or batch production towards repetitive manufacturing. The greater the degree to which the manufacturing system approaches repetitive manufacturing, the more relevant the Kanban technique.

13.4 Production activity control with Kanban

The Kanban system has been described as a *pull* system. We will now explore how this system works by taking the example of a very simple manufacturing and assembly system and illustrating the flow of kanban cards through it. The example is based on the two products, namely Stool A and Stool B, which were used to illustrate the logic of MRP in earlier chapters.

Under Kanban, only the final assembly line knows the requirements for the end product and, with this knowledge, it controls what is produced in the total manufacturing system using the following procedure:

> The final assembly line, having received the schedule, proceeds to withdraw the components necessary, at the times they are required and in the quantities they are required, from the feeding work centres or sub-assembly lines. These work centres or sub-assembly lines produce in lots just sufficient to replace the lots that have been removed. However, to do this, they also have to withdraw parts from their respective feeder stations in the quantities necessary. Thus, a chain reaction is initiated *up-stream*, with work centres only withdrawing the components that are required at the correct time and in the quantities required.

In this way, the flow of all material is synchronized to the rate at which material is used on the final assembly line. Amounts of inventory will be very small if a regular pattern exists in the schedule and if the deliveries are made in small quantities. Thus, Just in Time can be achieved without the use of controlling work orders for parts at each work centre.

13.5 The kanban card types

Kanban is the Japanese word for card. Kanbans usually are rectangular paper cards placed in transparent covers. There are two types of card mainly in use:

(1) **Withdrawal kanbans** Withdrawal kanbans define the quantity that the subsequent process should withdraw from the preceding work centre. Each card circulates between two work centres only – the user work centre for the part in question and the work centre which produces it.

(2) **Production kanbans** Production kanbans define the quantity of the specific part that the producing work centre should manufacture in order to replace those which have been removed.

Each standard container is assigned one of each card type. Examples of each card type are shown in Figure 13.4 (based on Monden 1983).

The withdrawal kanban, for example, details both the name of the consuming work centre and the work centre which supplies the part described by the item name and number on the card. The precise location in

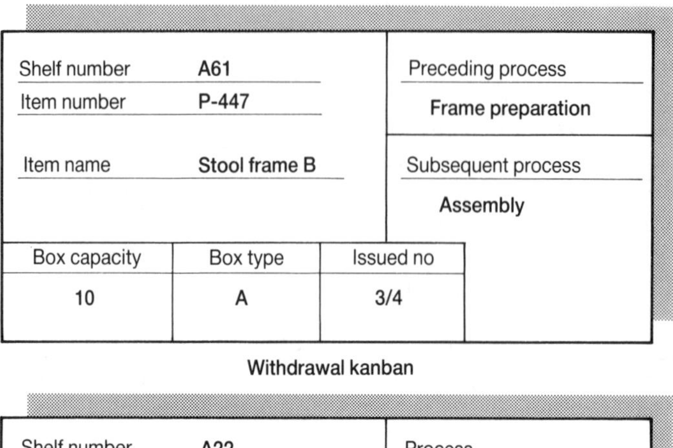

Figure 13.4 Kanban card types.

the inventory buffer is detailed, as well as the type of standard container used and its capacity. The issue number in the case shown in Figure 13.4 reveals that it is the third kanban issued out of four.

The production kanban details the producing work centre name, the part to be produced and where, precisely, in the buffer store it should be located.

There are other types of kanban differentiated by colour, shape or format, such as subcontract, emergency, special and signal cards. However, the two cards just described are the basic types used in the Kanban system.

13.6 The flow of kanban cards

To explain the flow of these cards through the production process, some examples will be taken from a simulation model of a Kanban controlled production and assembly system in whose development the authors were involved. This system consists of a main assembly line of five assembly stations. The line is a mixed model line which simulates the assembly of two different products. Each product passes through the assembly line where components are assembled to it at each assembly station, as illustrated in Figure 13.5. Each of these stations, in turn, is fed components by a feeder line which can include up to four work centres.

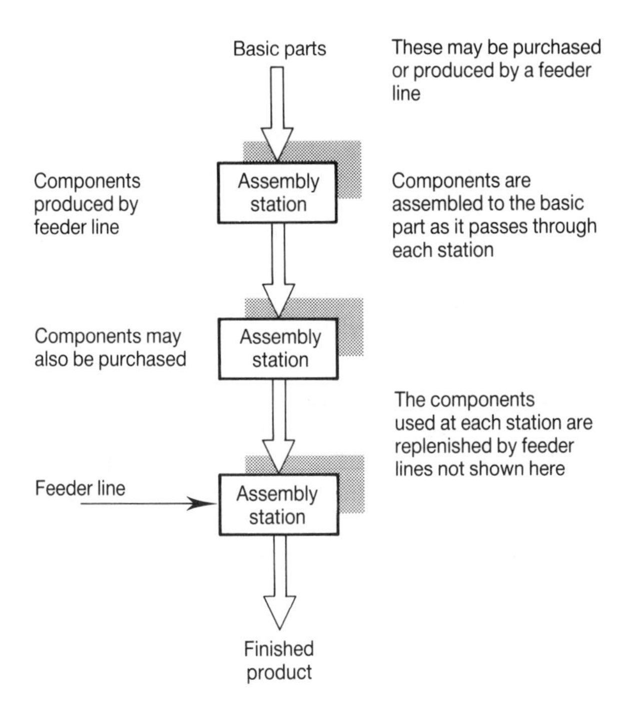

Figure 13.5 The production assembly line.

This model of the assembly line with its feeder lines will be used to show how production and withdrawal kanbans flow through the production system. The production process represented above highlights two different modes of material movement, namely:

(1) the flow of parts along an assembly line,

(2) the flow of parts along the feeder lines.

Section 13.7 will describe the flow of parts along the assembly line while Section 13.8 will be concerned with the flow of parts along the feeder lines.

13.7 The flow of parts along the assembly line

The most important aspect of the flow of material within the assembly line is that it is controlled *without* the aid of kanban cards. The material which forms the base product and onto which all components will be assembled arrives at the first station of the line. These parts may be produced by another feeder line or, alternately, may be purchased parts. The order in which these parts arrive is determined by the final assembly schedule. From the assembly schedule a cycle time is determined, within which all stations must have completed their respective operations on the part.

The flow of material is dictated by the cycle time of individual operations. The partially completed products progress to the following assembly station when the operation at the present station has been completed at the end of the cycle. However, if a problem arises at an assembly station and the operation has not been completed within the allowed cycle time, the whole assembly line stops production. This is signalled by the use of ANDON lights (see Chapter 12) which use colours to indicate the condition of the line. For example, a red light may mean that an assembly station has been unable to complete its work or an orange light might mean that there is not enough material available.

When such a situation arises, every operator in the vicinity will help to solve the problem and start the line moving again. This may involve a short term solution, such as expediting material to the station, and a longer term approach which determines why the material was not at the assembly station in the first place and which ensures that the problem does not occur again. Once the problem has been solved the light changes to green and production resumes.

Material is consumed at each station from an incoming material stock point which is fed by the appropriate feeder line (or storage point in the case of purchased items) and which has standard containers and associated withdrawal kanbans. Once the parts are removed from a container, the kanban is released to the feeder line and brought with an empty container to the work centre which feeds the assembly station. A full container is then brought back and placed in the incoming material store.

We now go on to look at the various types of manufacturing operations carried out at the feeder lines and how the material flow between these operations is controlled using the kanban cards.

13.8 Material movement in the feeder line

Before describing the flow of kanbans along the feeder lines, we will briefly examine the various classes of manufacturing operations which are carried out in a system, such as those illustrated in Figure 13.5.

Feeder lines are basically sequential processes in which a set of discrete functions are carried out on raw materials and semi-processed items. These different functions can be categorized into two main groups, namely materials handling functions and material processing operations. In materials handling, the material is simply moved from one location to another.

The material processing operations, in turn, can be subdivided into four distinct categories (see De *et al.* (1985), namely:

(1) Disjunctive processing operations.
(2) Locational processing operations.
(3) Sequential processing operations
(4) Combinative processing operations.

We will now briefly review each of these types of processing operation in turn.

(1) Disjunctive operations transform a single piece of raw material into many components. A punch press set-up to punch out a number of components from a single sheet of raw material is an example of a disjunctive processing operation. Another example is an automatic chucking lathe, designed to produce a number of components from a single piece of bar stock. The essential feature of a disjunctive operation is that a single piece of raw material results in a number of items produced – a one to many relationship.
(2) Locational operations involve the storage of material in a location for period of time. Warehousing is an example of a locational operation.
(3) Sequential operations occur where a raw material or a partially finished component is modified at a work centre and emerges intact as a single identifiable component. An example of a sequential operation is simple turning on a conventional lathe where, perhaps, the outside diameter of a component is turned down a specified value. The difference between a sequential operation and a disjunctive operation is that many items emerge from a single item input in a disjunctive operation, whereas an item is merely transformed or modified in a sequential operation.
(4) Combinative operations occur when a number of components or sub-assemblies are assembled, or joined together, to produce a single assembly or product. All assembly operations, by definition, are combinative. Consider, for example, the stuffing of a PCB (Printed Circuit Board) in the electronics assembly business. A series of components are assembled onto a bare PCB to produce a stuffed PCB.

The distinction between disjunctive, sequential and combinative processing operations is further developed in Figure 13.6.

Differing combinations of these four categories represent all possible types of discrete parts manufacturing. These four classes of manufacturing operation will be used to illustrate the use of kanban cards in controlling the flow of production on a feeder line in a JIT system. A pure assembly line consists totally of combinative processing operations, where components were added to the assembly as it progressed along the various assembly stations on the line.

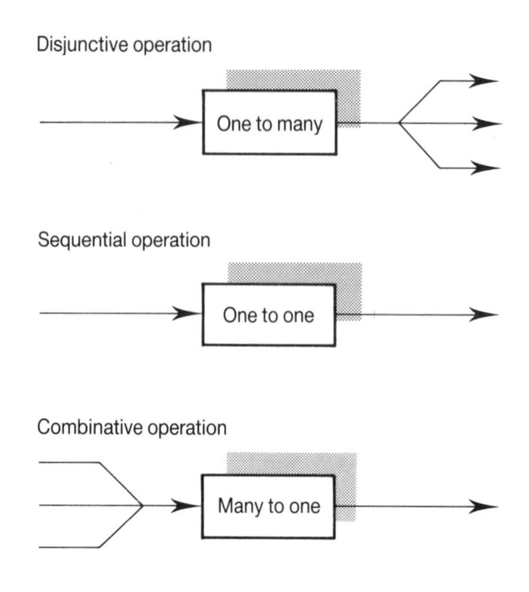

Figure 13.6 Types of manufacturing operations.

A typical feeder line can be composed of a number of work centres. Each work centre has two *stores* – one for incoming components and a second for components that have completed processing at that work centre. These stores could be considered locational processing operations. Some work centres are involved in combinative processing operations while others have sequential operations. At the beginning of the feeder line, the work centres are frequently disjunctive in nature since a single piece of raw material is processed to produce a number of components.

As an illustrative example of the nature of the work centres in the feeder lines, let us return to a *modified version* of the product that was used as an example to explain the workings of MRP in Chapter 5. A word of caution is in order here. This example is set up to facilitate the explanation of the operation of the kanban cards and, as such, is very simplistic.

Let us assume that Gizmo-Stools Inc. manufactures two types of stool, namely a four legged stool and a three legged stool. The original product structures are shown again in Figure 13.7.

In this case, let us further assume that Gizmo-Stools Inc. manufactures, as opposed to purchases, the legs of the stools. Let us also assume that the production system which manufactures and assembles the stools consists of a feeder line which produces the frame, another feeder line which produces the legs from tubular steel, and an assembly line. The frame is the basic item that enters the first assembly station where the legs are assembled to the frame to form the bare stool. The next assembly station produces the complete stool by adding a cushion to the bare stool.

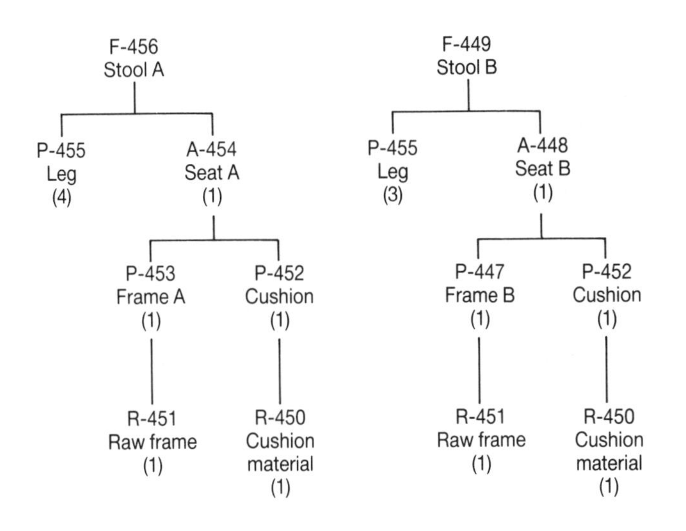

Figure 13.7 Two product structures.

For the purpose of illustrating the flow of material within a feeder line, we shall examine the line that produces the legs for the stools, where the legs are produced from lengths of tubular steel. Let us assume that there are three work centres in this line. The first cuts the tubular steel to length and produces the legs in groups of either three or four. The next work centre forms the tops of these legs to facilitate their joining to the stool frame. At the final work centre on this feeder line, rubber feet are attached to the bottom of the legs. Let us call these work centres, cutting, forming and finishing, respectively.

The legs are produced therefore by the work centres in the feeder line according to the product/process structure diagram in Figure 13.8. This product/process structure shows that, at different work centres, each of the four basic categories of material processing operations (locational aside) identified above are carried out in the feeder line.

13.8.1 The flow of cards along the feeder line

This section reviews the movement of material between work centres and indicates how the kanbans are used to control it. The above example will be used to show the movement between the forming and cutting work centres on the feeder line. These operations are mainly of a material handling nature where newly cut legs are moved from the cutting work centre to the forming work centre. Figure 13.9 will be used as a basis for this description.

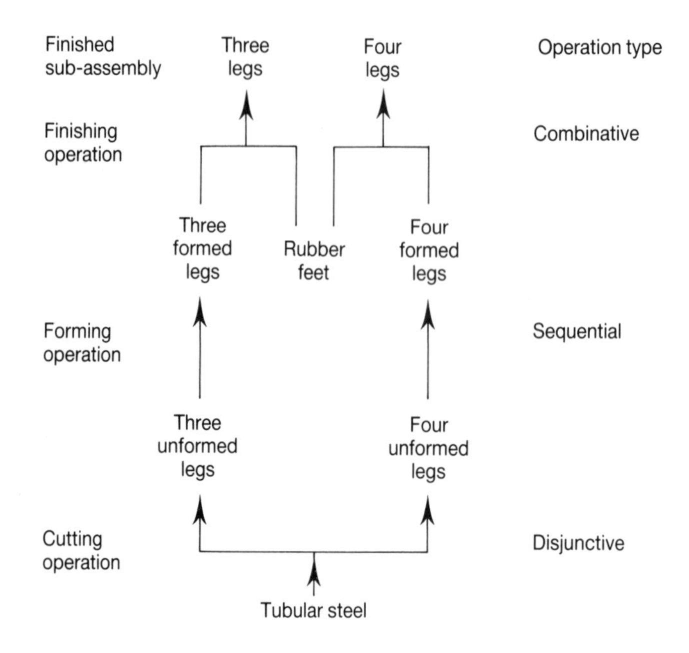

Figure 13.8 Manufacture of legs.

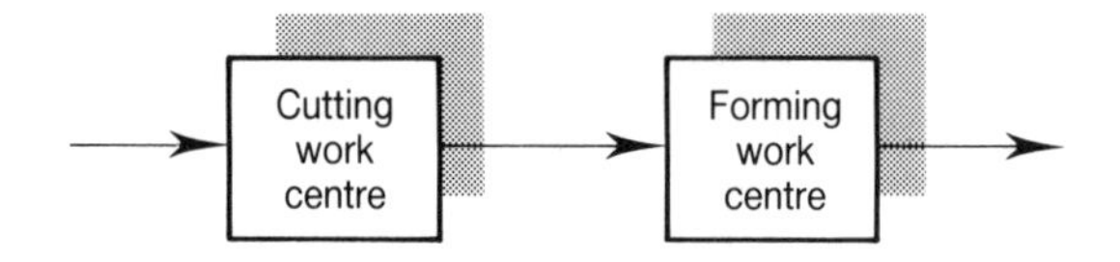

Figure 13.9 Flow of material between work centres.

Let us assume that the forming work centre has a requirement for material, i.e. preformed legs. These are the outgoing components from the cutting work centre. Both work centres have two stockpoints – one for incoming material and a second for outgoing material. Stocks of parts are stored in standard containers which are located at these stockpoints. There are two types of standard container – one for batches of three legs and a second for batches of four legs.

Each standard container has a corresponding kanban card which details the type of material in the container and the quantity of that material. Those containers in the incoming material stockpoint are associated with withdrawal kanbans while those in the outgoing material stockpoint are linked to production kanbans.

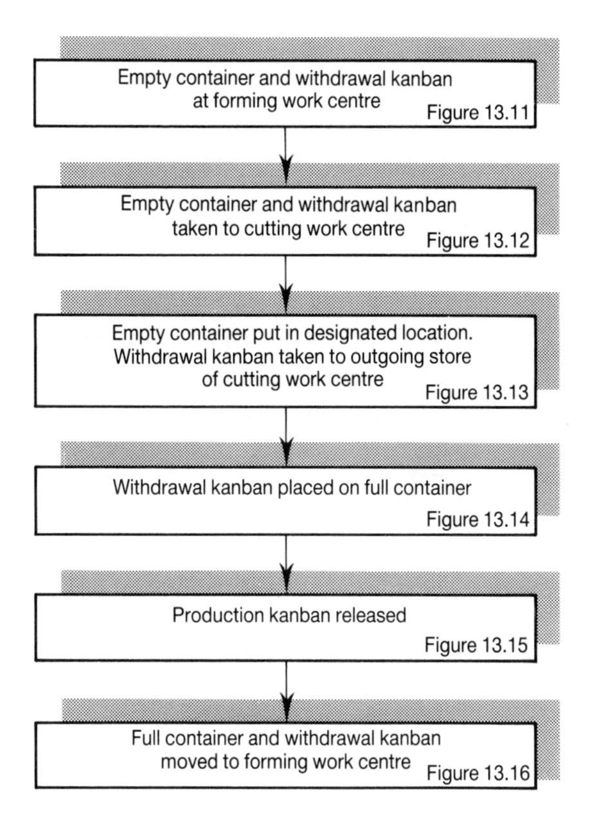

Figure 13.10 Events in moving materials between work centres.

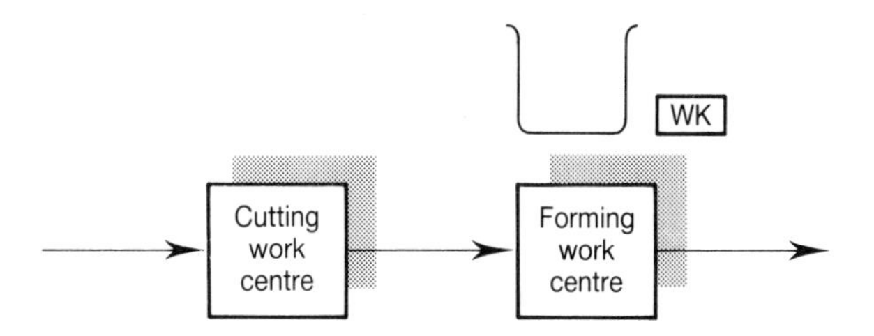

Figure 13.11 Empty container and withdrawal kanban at forming work centre.

For the forming work centre to require material, there must be an empty standard container and a corresponding withdrawal kanban. The flow chart in Figure 13.10 depicts the various steps in the movement of material and kanbans. Each step is depicted by a corresponding figure described later.

The empty container and the withdrawal kanban signal the requirement for materials at the forming work centre, as shown in Figure 13.11.

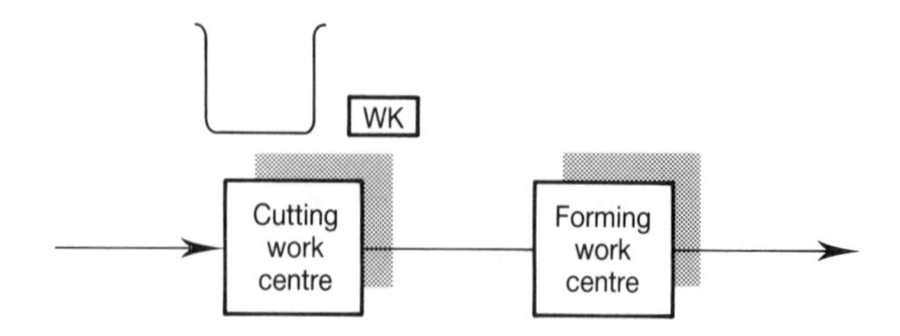

Figure 13.12 Empty container and withdrawal kanban at cutting work centre.

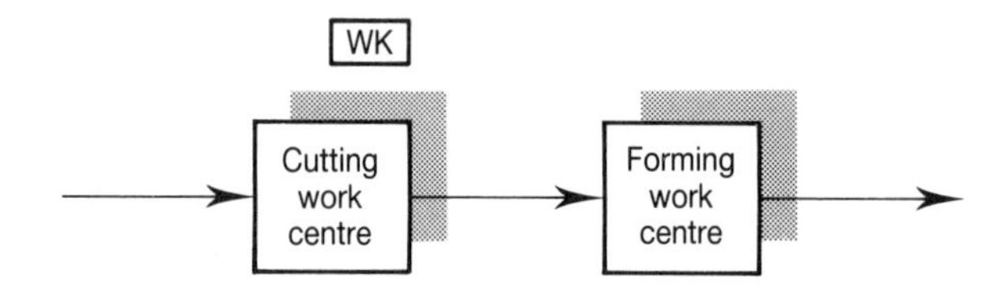

Figure 13.13 Withdrawal kanban taken to outgoing store.

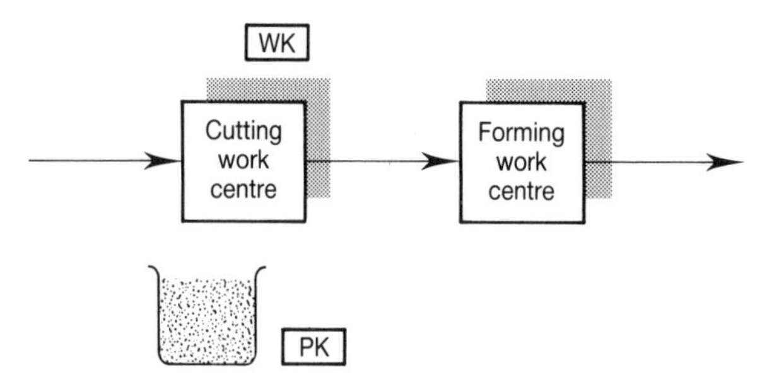

Figure 13.14 Container of required material selected.

Figure 13.12 shows this empty container and withdrawal kanban being taken to the outgoing material stockpoint of the cutting work centre.

At the cutting work centre, the empty containers are placed in a designated location and the withdrawal kanbans are brought to the outgoing material stockpoint, as shown in Figure 13.13.

At the outgoing material stockpoint, a standard container which has the same parts as specified on the withdrawal kanban is removed, as shown in Figure 13.14. Remember that the outgoing material of the cutting work centre is the same as the incoming material of the forming work centre.

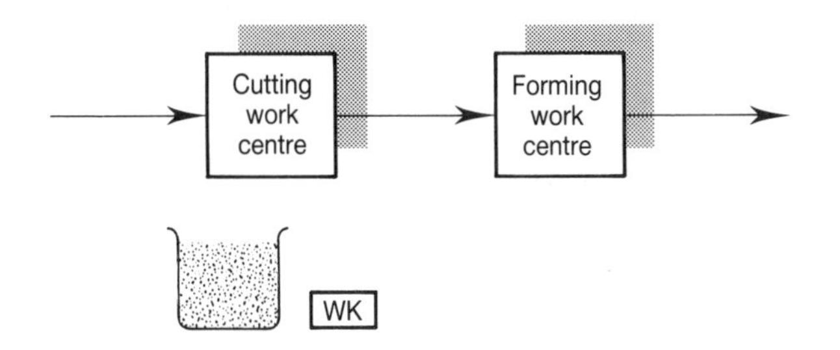

Figure 13.15 Production kanban released.

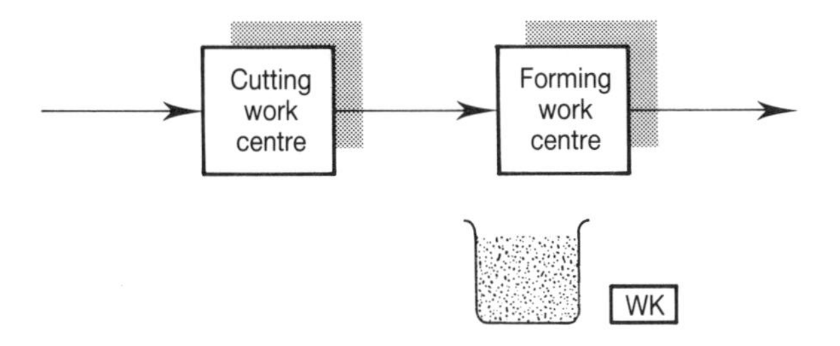

Figure 13.16 Full container and withdrawal kanban moved to forming work centre.

Attached to these containers are the production kanbans and a quick check will verify that the parts are those that are required. The kanbans are then switched – the withdrawal kanban being attached to the container and the freed production kanban placed in a designated location as in Figure 13.15.

Figure 13.16 shows the full container with the attached withdrawal kanban brought to the incoming material stockpoint of the forming work centre.

The flow of material between work centres has now been fully described, showing how it is controlled using withdrawal kanban cards. Section 13.7 will describe the flow of material and kanbans within an individual work centre.

13.8.2 The flow of cards within a work centre

Each work centre has an inbound and an outbound stockpoint, as is illustrated in Figure 13.17. The material flows from the preceding work centre into the inbound stockpoint, as discussed above for the case of the

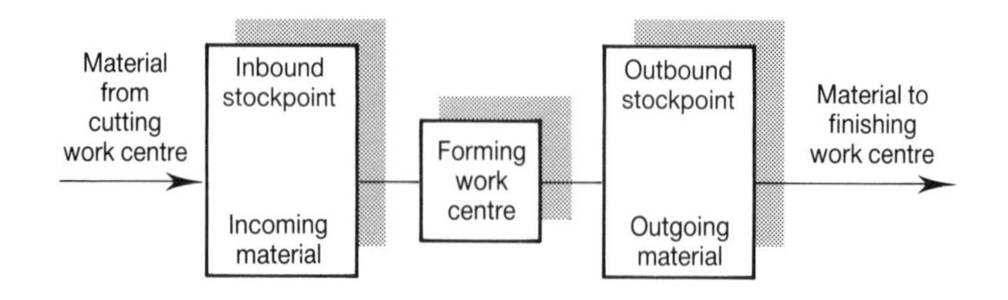

Figure 13.17 Schematic representation of work centre.

cutting and forming work centres. This is stored until the forming work centre requires material. The outgoing product at this work centre then proceeds to the outbound stockpoint. From here it flows to the next work centre. These work centres can be any of the three material processing operations – disjunctive, sequential or combinative. For illustrative purposes let us use, as an example, the forming work centre. It represents a sequential processing operation where incoming material is simply modified. The other work centres in the line represent both of the other processing operations. The cutting work centre performs a disjunctive operation where tubular steel is cut to form legs. On the other hand, the finishing work centre represents a combinative operation, where the rubber feet and legs are assembled to produce the finished legs.

Figure 13.18 is a flow chart representing all the steps associated with the flow of kanbans within this work centre. Each step is depicted by a corresponding figure described later.

The empty container with the withdrawal kanban is brought from the incoming material store of the finishing work centre to the outgoing material store of this, the forming work centre. As shown in Figure 13.19, the empty container is left in a designated location.

A full container of the parts described (i.e. three or four formed legs) on the withdrawal kanban is located in the outgoing material store. This has a production kanban attached, which contains similar information to that on the withdrawal kanban. The production kanban is removed to a designated location and the withdrawal kanban attached to the full container as in Figure 13.20.

This full container and the withdrawal kanban are then brought to the incoming material store of the finishing work centre, as illustrated in Figure 13.21. As production kanbans become available, they are moved to the incoming material store of the forming work centre.

The production kanban also contains information about the materials required to produce the outgoing goods. This will specify whether three or four legs are required. Figure 13.22 shows a full container of these materials being removed from the store and the withdrawal kanban being released.

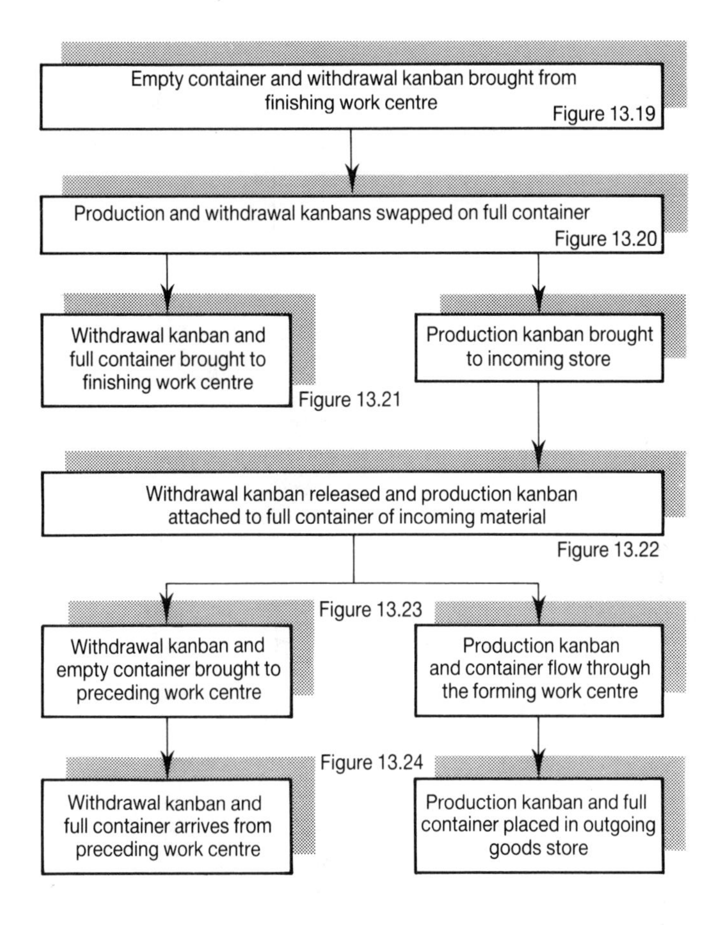

Figure 13.18 Events in moving material within a work centre.

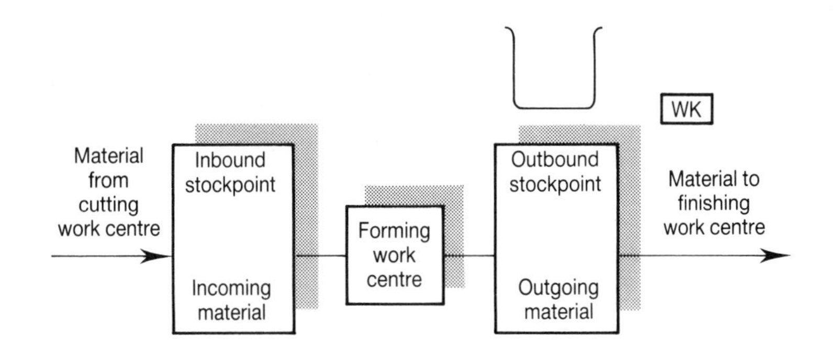

Figure 13.19 Empty container and withdrawal kanban brought from finishing work centre.

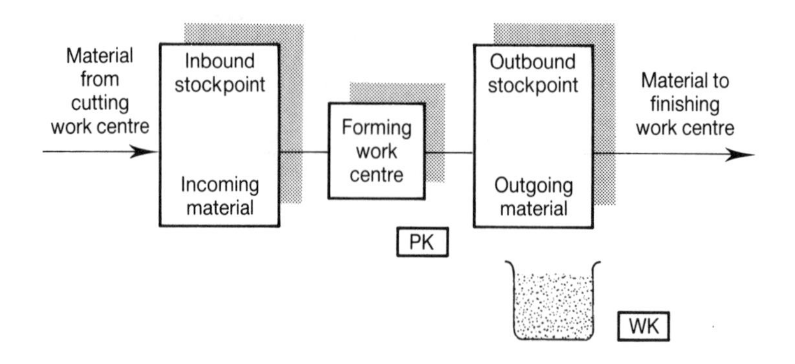

Figure 13.20 Production and withdrawal kanbans swapped on full container.

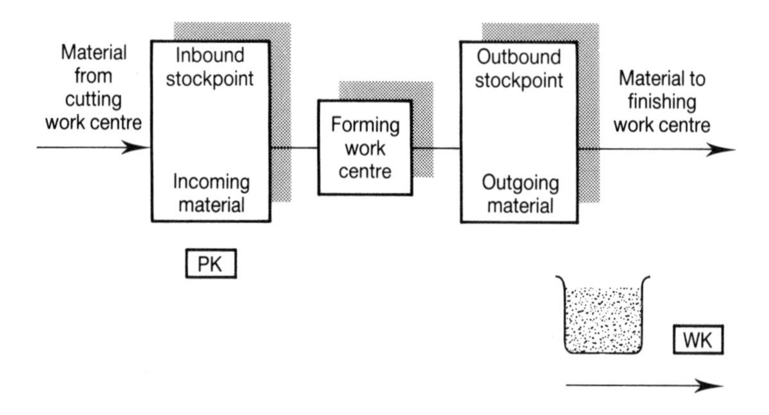

Figure 13.21 Withdrawal kanban and full container brought to finishing work centre.

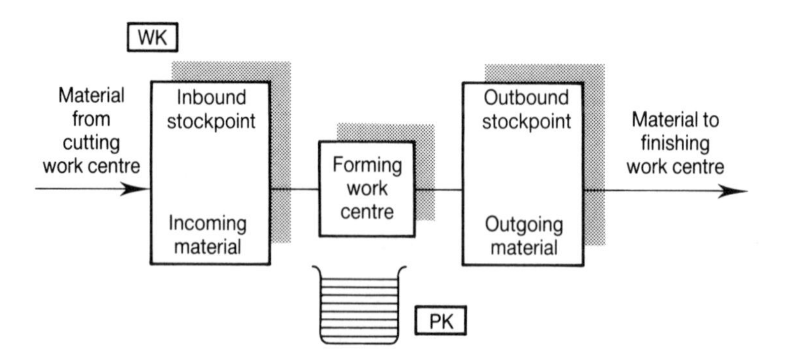

Figure 13.22 Withdrawal kanban released and production started.

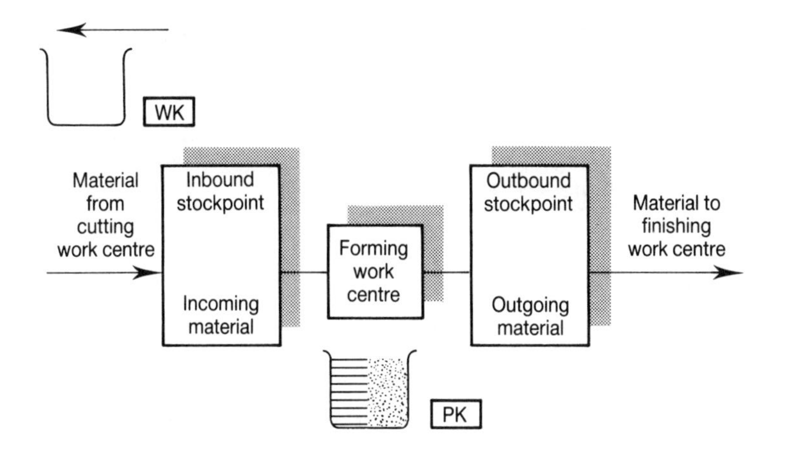

Figure 13.23 Processing partially complete.

If the material processing operation is a combinative operation, as in the finishing work centre, the production kanban details that a container of legs and rubber feet are required. However, for a disjunctive operation (as in the cutting work centre), production kanbans for three and four legs will initiate the release of the same withdrawal kanban since both involve the same raw material, namely tubular steel.

Once an empty container is available, it and the withdrawal kanban are brought to the outgoing material store of the cutting work centre to obtain a container of incoming material. Meanwhile, the production kanban stays with the container of material right through the production process at the forming work centre, as shown in Figure 13.23.

When the preformed legs have been processed into formed legs, the container and the production kanban are placed in the outgoing material store. At some point during the processing of the legs, a full container of unformed legs is brought from the cutting work centre outgoing store with an attached withdrawal kanban and placed in the incoming material store as in Figure 13.24.

13.9 Kanban card usage

It is clear that for a Kanban system to operate effectively, very strict discipline is required. This discipline relates to the usage of the kanban cards. This requirement for discipline also serves to illustrate the need for well documented manufacturing procedures and a well trained group of operators who are aware of the procedures and who are motivated to follow them

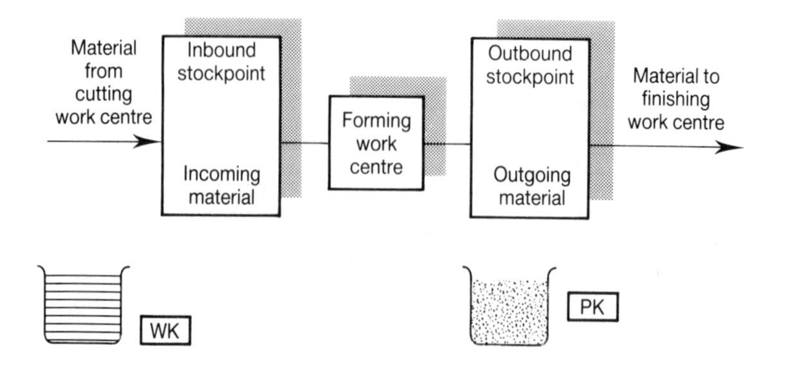

Figure 13.24 Full containers of incoming and outgoing material at stockpoints.

rigorously – a confidence born of experience of good operator practice. There are five guidelines on the usage of the kanbans which help achieve JIT production:

(1) A work centre should withdraw only the items which it requires from the preceding work centre in the quantities required and, equally importantly, at the required time.

   There are a number of operating principles which support this:

   - No removal of material is allowed without an available withdrawal kanban and an available empty container.
   - A withdrawal of more parts than indicated on the withdrawal kanban is not allowed.
   - The kanban must be attached to one of the items within the container.

(2) A work centre or process should only produce those items which have been removed by the following work centre or process.

   The freed production kanbans act as a *schedule* for the work centre. The work centre is not allowed to produce greater quantities than stipulated on the production kanban and the sequence of operations in the work centre or process must follow the sequence in which the production kanbans were freed. As production is initiated by the final assembly schedule released to the assembly line, the schedule is passed back through the production system by the release of production kanbans.

   Rigorously adhering to the above guidelines results in what is effectively an *invisible* conveyor line, constructed and controlled by the flow of kanbans through the production system.

(3) Defective or substandard items should never be passed to a following work centre. This implies rigorous quality control at each work centre or

step in the production process. Allowing defective parts to stay within the production system will greatly upset the flow of parts at a later stage when the defective part is detected.

(4) The level of inventories in the production system is dictated by the number of kanbans since each kanban represents the contents of a standard container. The number of kanbans should therefore be minimized. By reducing the number of kanbans and the size of each container the level of inventories is progressively reduced.

(5) The Kanban system is only suitable for dealing with relatively small fluctuations in the demand pattern in the final assembly line. The system is only relevant to a repetitive manufacturing situation and large changes in demand cannot be accommodated within it.

Sudden large changes in the demand for products should have been eliminated using the JIT techniques outlined in Chapter 12, i.e. by following the JIT approach to the market and design of products and by using a manufacturing system which facilitates flow based production. Clearly, if it is not possible to arrive at a stable master schedule for the end level items in the bill of materials, the Kanban system cannot be used. Small fluctuations in the demand can be handled by increasing the circulation frequency of the kanbans, by increasing overtime or by hiring temporary operators.

The effort involved in making the Kanban system work well is tremendous. It requires extensive use of the JIT techniques outlined in Chapter 12 to establish the correct environment for Kanban and maintenance of the discipline of the kanban cards at all times.

13.10 The full work system – Kanban for automation

How can the Kanban system be used with automated manufacturing equipment? A variant of the system, known as the *full work system*, is used. With automated manufacturing processes, as with manual operations, it is necessary to match the number of items produced with those withdrawn. (See automatic production level controls in Chapter 12.) It is necessary to match the capacities and speeds of production of different machines, otherwise a build-up of inventory will arise between machines. Also, it is necessary for each machine to be sensitive to any problems on the following machine which might result in it being unable to process further parts. These problems are resolved by using the full work system. This system is illustrated for two machines in Figure 13.25 (adapted from Monden (1983)).

The two machines are connected by a magazine or chute to store the items completed at machine A. The inventory level can be set to, say, 4 units. When machine A has processed 4 units, a limit switch (connected to the magazine) is activated and machine A stops processing. Machine B

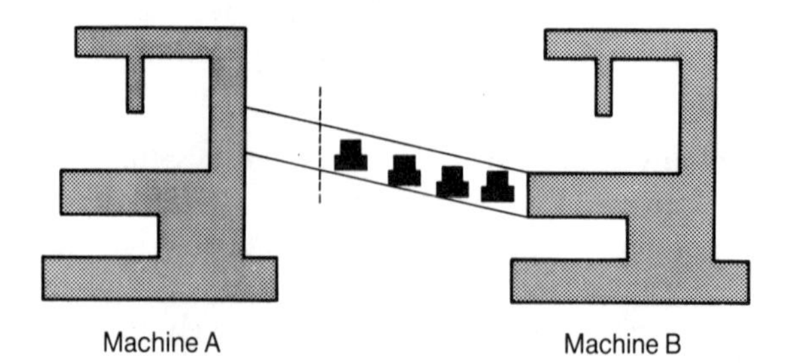

Figure 13.25 Full work system.

continues to withdraw the parts until there are, say, 2 units in the magazine, then another switch is activated and machine A starts processing again. In this way, only a fixed quantity of work in progress is allowed at each process, and processes are linked, thus preventing unnecessary processing and build-up of inventory in the preceding processes.

The similarity between this system in its use of limit switches and a kanban card has resulted in the switch sometimes being referred to as an *electric kanban*.

13.11 The single card Kanban

So far, we have been discussing a Kanban system using two types of kanban cards, namely the withdrawal and production cards, i.e. a two card kanban system. However, there are other kanban systems in existence that use only one type of kanban – the withdrawal kanban – i.e. a single card system.

The single card system is sometimes used as a first stage in the development of a full Kanban or two card system. The operation of such a single card system will now briefly be described.

The procedures in a single card kanban system are relatively simple. Each production centre produces parts according to a daily schedule, which is determined from the final assembly schedule along the lines of an MRP system. When the production centres require component parts, the withdrawal kanbans are used to acquire the required parts from other work centres.

The procedure is as follows:

- As each container of incoming parts is started at a work centre, the associated withdrawal kanban is placed in a specific location. These incoming parts are then processed.
- At regular intervals, the withdrawal kanbans are collected. These kanbans contain information detailing the part type and work centre

which produces these parts. A full container is removed from the producing work centre, the withdrawal kanban attached and the container is returned to the incoming parts store of the consuming work centre.

Some points which differentiate this system from the two card system are:

(1) The inbound stockpoint is not so important, given that deliveries of parts to the work centre are controlled very tightly, so there is always a minimum number of containers available to be processed. This relieves space and prevents confusion around a stockpoint.

(2) Processed parts are allowed to build up at the outbound stockpoints. This may not be a serious problem if the production planners can readily associate the required quantities and timings of parts with the final assembly schedule. In such a system, the withdrawals of required component parts are controlled, while the production is driven by the daily parts schedule.

(3) It should be noted that this single card kanban system is very similar to the *two bin* reorder point system. The difference primarily relates to the environment in which it is applied. The single card Kanban system is used in an environment of standard container and flow based production systems, and all of the other attributes of a JIT environment. Also, only a small number of containers are allowed at work centres and, since the quantity per container is usually small, many replenishments are required per day.

    The single card system could be characterized as a *push* system for production control and a *pull* system for material delivery, since the cards are used only to pull materials between work centres.

13.12 Relationship to vendors

If the flow of kanban cards back through the production process is followed logically, it is clear that, ultimately, the incoming raw material and the purchased parts point is reached. This leads to the question of how the Kanban system might be extended to outside vendors. On the one hand, we can have large inventories of each part which are replenished by suppliers at weekly, or perhaps longer, intervals. However, this defeats one of the essential purposes of JIT which is to reduce inventories. On the other hand, we can carry the Kanban process right out and into the vendors' production systems. This procedure normally involves regular and frequent deliveries from the vendors and is achieved, as indicated in Chapter 11, through the establishment of close cooperation with suppliers and the sharing of as much information as necessary to help the suppliers' organizations achieve a JIT system. In effect, it involves establishing a true partnership with trusted suppliers.

13.13 Kanban as a productivity improvement technique

If the procedures of Kanban outlined earlier are followed rigorously, the level of work in progress inventory can be controlled by the number of cards issued for each component part into the system. This is because each card corresponds to one standard container and knowing the size of the container and the number of cards on the floor a simple calculation gives the inventory level.

Therefore, by reducing the number of cards issued for a particular part, the process inventory level for that part falls. Eventually, if the levels are reduced low enough, a work centre will run out of material and stop processing. This stops the whole line and major efforts are made to get it running at the lower level of inventory, by either increasing the number of operators, reducing set-up times or redesigning processes.

The thinking is as follows. Unnecessarily high inventory serves to disguise inherent problems and sources of inefficiency in the production process. Through gradual reduction of inventory levels, production problems are highlighted and are progressively eliminated.

13.14 Assumptions necessary for Kanban to work

As indicated above, strict rules must be followed to enable the Kanban system to work efficiently and effectively. Corresponding to these rules are some fundamental assumptions about the nature of the manufacturing system, within which the Kanban system operates. We will now briefly revisit these assumptions.

Since each daily assembly schedule must be very similar to all other daily schedules, it is essential that it is possible to *freeze* the master production schedule for a *fixed time period*, possibly of at least one month. The final assembly schedule must also be very level and stable. Any major deviations will cause a ripple effect through the production system causing *up-stream* work centres to hold larger inventory stocks. What is required ultimately is that the manufacturing system conform as closely as possible to the *repetitive manufacturing* system model outlined earlier in this chapter.

To run a mixed model system effectively, requires mixed model *capability* in *all stages* of the production process. Mixed model manufacturing and assembly, as seen in Chapter 12, involves frequent changes and set-up at the individual work centres. It follows that the set-up times in all work centres be as small as possible and that the set-up procedures be continuously reviewed to this end. A logical conclusion from this is the need for balancing between all operations in order to synchronize the starting and ending of work routines. This ensures that parts are fed to the assembly line at the same rate as they are consumed.

The plant configuration and layout must be *flow based* in order to link all operations to the final assembly line, thus reducing transport times and highlighting interdependencies between work centres. Each work centre should have inbound and outbound stockpoints and there should be a fixed flow routing for each part through the work centres. The system should be designed to operate at somewhat less than full capacity in order to provide some flexibility if, and when, problems arise.

A *multiskilled* workforce is a basic requirement for Kanban. Such a workforce can be flexible in its work organization in that it can handle many different machines and operations. Moving people around allows supervisors to change the capacity of individual work centres. It is sometimes difficult to secure the level of operator commitment and involvement that is required for such activities and, in fact, many non-Japanese companies consider this a major obstacle to the installation of a strict Kanban system in their manufacturing plants.

13.15 Conclusion

This chapter has discussed the Kanban system and presented it in the context of JIT thinking.

There are clearly some limitations and disadvantages to the Kanban system. Kanban is intrinsically a system for repetitive manufacturing. It will not succeed without modification in a non-repetitive manufacturing environment.

As mentioned above, Kanban requires a levelled schedule, standard containers and very strict discipline. It could be considered inflexible in that it cannot easily respond to irregular changes or to large unexpected changes in market demand and it clearly requires great cooperation from outside suppliers.

From the perspective of the process, it places emphasis on process technologies, such as product based flow configurations, and may therefore require considerable investment in developing new methods, procedures, jigs and fixtures, etc., perhaps even new capital equipment.

If well implemented, there are many advantages to the Kanban system. The most important of these are that it stimulates productivity improvement, reduces inventory and production lead time, and, within the constraints of the product design and the manufacturing system design, allows the plant to respond to predictable small market variations. Kanban is a simple system of flow control with visible means of inventory control, which is simple to understand. It involves very little paperwork compared with other systems – such as the shop floor control modules of typical MRP systems – and is able to set valid priorities.

Improvements in operations are promoted through the development of a plant layout (which facilitates a smooth production flow and the redesign of equipment), jigs and tools for fast set-ups and balanced production rates. Other benefits arise from the training of the workforce to be multiskilled, greatly reduced scrap rates, higher quality levels and saving on space due to lower inventory levels requiring less physical storage space.

Evolution towards more efficient repetitive manufacturing is ensured by promoting the reduction of set-up time and lot size, and the development of formal stockpoints within each work centre.

To summarize, this chapter has described the shop floor control system, Kanban, which is an integral part of the realization of the just in time philosophy in repetitive manufacturing environments. Its operation has been described with particular emphasis on how the flow of material is controlled through production and withdrawal kanbans.

In our view, Kanban is the least important aspect of JIT. What is important is that, by implementing the ideas described in Chapters 11 and 12, it may be possible to streamline manufacturing systems, thereby realizing huge benefits in terms of productivity, efficiency and flexibility. In some circumstances, the result of applying JIT may be such as to create the conditions for Kanban to be appropriate. Perhaps, in the majority of cases, this will not be so. Nevertheless, the benefits of implementing JIT are likely to be tremendous.

References

APICS. 1984. *JIT and MRPII: Partners in Manufacturing Strategy*, Report on 27th Annual APICS Conference on Modern Materials Handling, December, 58–60.

Boothroyd, G. and Dewhurst, P. 1982. *Design for Assembly: A Designer's Handbook*. Amherst, Massachusetts: University of Massachusetts.

Boothroyd, G. and Dewhurst, P. 1983. 'Design for assembly: choosing the right method', *Machine Design*, November, 94–98.

Browne, J., Furgac, I., Felsing, W., Deutschlaender, A. and O'Gorman, P. 1985. 'Product design for small parts assembly', in *Robotic Assembly*, edited by K. Rathmill. UK: IFS Publications Ltd, 139–156.

Burbidge, J.L. 1963. 'Production flow analysis', *The Production Engineer*, 42.

Burbidge, J.L. 1975. *The Introduction of Group Technology*. New York: John Wiley and Sons.

Cortes-Comerer, N. 1986. 'JIT is made to order', *IEEE Spectrum*, September, 57–62.

De, S., Nof, S.Y. and Whinston, A.B. 1985. 'Decision support in computer integrated manufacturing', *Decision Support Systems*, **1**, 35–56.

Edwards, J.N. 1983. 'MRP and Kanban, American Style', in *APICS 26th Annual International Conference Proceedings*, 586–603.

Gallagher, C.C. and Knight, W.A. 1973. '*Group Technology*'. London: Butterworths.

Groover, M.P. 1980. *Automation, Production Systems and Computer-Aided Manufacturing*. USA: Prentice Hall Inc.

Hall, R.W. 1981. *Driving the Productivity Machine Production Planning and Control in Japan*. USA: American Production and Inventory Control Society.

Hall, R.W. 1983. *Zero Inventories*. USA: Dow Jones–Irwin.

Ham, I. 1976. *Introduction to Group Technology*. Dearborn, Michigan: Society of Manufacturing Engineers, Technical Report MMR76–03.

Harrington, J. 1973. *Computer Integrated Manufacturing*. New York: Industrial Press, Inc.

Heard, E. and Plossl, G. 1984. 'Lead times revisited', *Production and Inventory Management*, third quarter, 32–47.

Hyer, N.L. and Wemmerlov, U. 1982. 'MRP/GT: a framework for production planning and control of cellular manufacturing', *Decision Sciences*, **13**, 681–700.

Laszcz, J.Z. 1985. 'Product design for robotic and automatic assembly', in *Robotic Assembly*, edited by K. Rathmill. UK: IFS Publications Ltd.

Lewis, F.A. 1986. 'Statistics aid planning for JIT production', *Chartered Mechanical Engineer*, June, 27–30.

Menga, G. 1987. (Private communication.)

Monden, Y. 1981. 'Adaptable Kanban system helps Toyota maintain Just in Time production', *Industrial Engineering*, May, 29–46.

Monden, Y. 1983. *Toyota Production System: Practical Approach to Production Management*. American Institute of Industrial Engineers.

Schneidermann, A.M. 1986. 'Optimum quality costs and zero defects: are they contradictory concepts' *Quality Progress*, November, 28–31.

Schonberger, R.J. 1982. *Japanese Manufacturing Techniques: Nine Hidden Lessons in Simplicity*. New York: The Free Press.

Schonberger, R. J. 1984. 'Just in Time production systems: replacing complexity with simplicity in manufacturing management' *Industrial Engineering*, **16**(10), 52–63.

Shingo, S. 1985. *A Revolution in Manufacturing: The SMED System*. USA: The Productivity Press.

Skinner, W. 1974. 'The focused factory', *Harvard Business Review*, May–June.

Treer, K.R. 1979. *Automated Assembly*. Dearborn, Michigan: Society of Manufacturing Engineers.

Wild, R. 1984. *Operations and Production Management Principles and Techniques*, London: Holt, Rinehart and Winston.

PART IV

Optimized production technology†

Overview

The major intent of this portion of the book is to introduce and review the relatively new production management system, OPT – an acronym for Optimized Production Technology. This system, which has been developed within the last decade, has two major components, namely, a philosophy which underpins the working system and a software package that produces manufacturing schedules through the application of this philosophy to the manufacturing system.

The OPT system has generated some interesting discussion, one might even say controversy, in the literature. In our view this controversy arises from two sources. Firstly, the term *optimal* has a strict scientific meaning and it is fairly clear that OPT is not optimal in the scientific sense. Secondly, the original information on OPT talked about a *black box* secret algorithm for generating schedules. This algorithm has never been made public to our knowledge. Some authors take the view that OPT is a competitor for MRP II and JIT. Others, notably Vollmann (1986), consider OPT as an enhancement to MRP II.

It has been suggested that the major tenets of the OPT philosophy can be applied to the management of a manufacturing system without recourse to the software package. Here we shall consider these tenets, or rules as they are called, with respect to their applicability.

This part is organized into two chapters. In the first of these we cover the thinking behind OPT and in the second we outline the operation of the software package as we understand it. We shall support this second chapter with an example similar to that used to outline the basics of MRP in Part II.

In Chapter 14, we consider the backbone of OPT – its *philosophy*. The OPT philosophy is founded on the basic assumption that the primary goal of any manufacturing business is *to make money*. OPT addresses itself to various aspects of manufacturing, from both the production and business

† OPT® is a registered trademark of Scheduling Technology Group Ltd.

perspectives. For example, scheduling, resource utilization and cost accounting are all important elements of OPT. The OPT philosophy can be condensed into *rules* which must be followed to attain the primary goal of making money. Application of these rules on their own will, it is claimed, improve the overall performance of a business.

In Chapter 15 we describe how the OPT software package operates, or rather, as much of the system as can be understood from reading available documentation. The system has, as inputs, complete descriptions of the facility, the products and the orders. OPT then produces schedules which must be followed exactly to achieve the gains that are attributed to this system. We shall complement this description of the system by manually attempting to simulate the operation of the OPT system.

In these chapters we will show that OPT is an important and viable approach to production management with an, as yet, unproven track record. OPT will, in the near future, be used primarily as a scheduling tool and its future as a complete production management system will likely be through interaction with modules of MRP II. As mentioned in the overview of Part II, OPT will form part of the new hybrid production management environments, complementing the other production management paradigms that have been presented. A final comparative review of the alternative production management philosophies will be made in Part V.

CHAPTER FOURTEEN

Optimized production technology philosophy

14.1 Introduction

In response to the continued success of Japanese manufacturing, a new approach to the management of manufacturing has been developed in the West within the last ten years. The OPT (Optimized Production Technology) approach contains many of the insights which underlie the Japanese Kanban system (see Part III) – an important element in just in time manufacturing.

14.2 Background to OPT

'The OPT philosophy contends that improving productivity is any step that takes the company closer to its goal' (Fox (1982a)). From the OPT perspective there is one, and only one, goal for a manufacturing company – **to make money**. All activities in the business are but means to achieve this goal.

This goal can be represented by three **bottom-line** financial measurements as follows:

(1) net profit,
(2) return on investment, and
(3) cash flow.

If the business takes actions that increase each of these simultaneously then it is moving in the right direction – towards the goal of making money. From the operational point of view, OPT defines three important criteria that are useful in evaluating manufacturing progress towards this goal. These criteria are throughput, inventory and operating expenses and are defined in the following way:

(1) **Throughput** Throughput is the rate at which the manufacturing business generates money through *selling* finished goods. It is *not* a measure of production. For example, in an automobile production plant, the manufacture of components is unimportant from the perspective of throughput as defined by OPT, since components, in general – ignoring the manufacture of spares for the purposes of the example – can only generate money for the business when they appear as part of a finished product. To generate money, finished automobiles must be sold. Therefore, throughput in the OPT sense is concerned only with the rate at which finished units are sold.

(2) **Inventory** Inventory is defined by OPT to be the raw materials, components and finished goods that have been paid for by the business but have not, as yet, been sold. Inventory thus excludes the added value of labour and overhead in order to eliminate the distortions that may be caused by accounting profit and losses into inventory. For example, in Chapter 5, the legs, frames and cushion material used to manufacture the stools in Gizmo-Stools Inc. plus the finished stools in stock at any moment, are all considered inventory. These are materials that have been paid for and have not yet been converted back into money through the sale of finished products.

(3) **Operating expenses** Operating expenses are defined as the cost of converting inventory into throughput. Operating expenses include the cost of direct and indirect labour, heat, light, etc. and production facilities, to mention but a few. In the example of Gizmo-Stools Inc. above, the operating expenses are a measure of the costs incurred to produce the seat of the stool, assemble the legs to the seat, paint the stool, etc. These expenses have to be recovered by the sale of finished stools.

Changes in any of these three elements, such as increasing the throughput or reducing the inventory level, result in changes in the *bottom-line* financial measurements listed above. The effects that changes in throughput, inventory or operating expenses have on the financial measurements of net profit, return on investment and cash flow will now be considered.

- An *increase* solely in throughput means a simultaneous increase in net profit, return on investments and cash flow. The business is selling more finished goods while maintaining a stable level of inventory and operating expenses. This obviously means a greater influx of money, larger profits and earlier recovery of investments.
- A similar result applies with a *reduction* in operating expenses. In this case the cost of producing the finished product is reduced while the

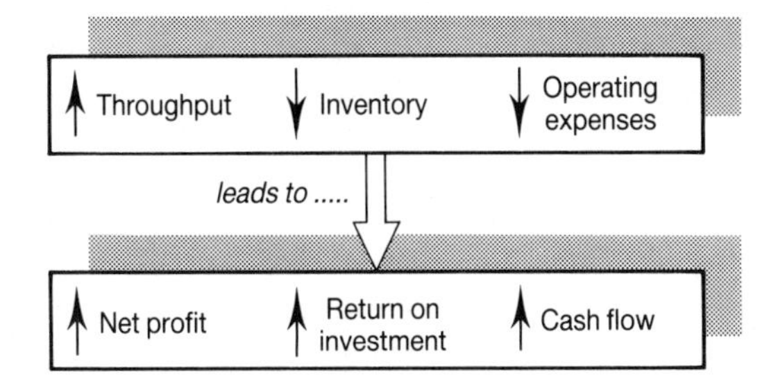

Figure 14.1 Operation measures and financial measures.

- inventory and the rate at which the finished products are sold remains unchanged. It is clear that this increases cash flow, net profit and return on investment.
- A *reduction* in inventory directly impacts return on investment and the cash flow. Here, less costs are associated with a particular period since the inventory is lower. This improves the flow of money and helps recover investment. Profit is not changed, since the cost of raw material has not changed and neither has the cost of transforming the raw material into a finished product.

Therefore, the goal of manufacturing, as seen by OPT, is to increase throughput while *simultaneously* decreasing inventory and operating expenses, as shown in Figure 14.1.

OPT, which was initially developed in Israel during the 1970s, is an analytical technique that is designed to achieve this goal of increasing throughput and decreasing inventory and operating expenses through realistic *optimized* schedules. Closely coupled with the analytical technique is the OPT philosophy. This philosophy basically consists of ten rules. These rules and the analytical technique have been computerized to give a software product called OPT.

The ten rules of OPT, which form the basis of the OPT approach, may be applied to the manufacturing organization without recourse to the software system. Therefore, the rest of this chapter will be devoted to describing and discussing these rules. Chapter 15 will concentrate on the software product, OPT, and try to illustrate how it operates using an example.

14.3 The OPT philosophy

The OPT philosophy rests on the premise that the primary goal of manufacturing is to make money. To attain this goal, OPT considers activities on the shop floor to be critical. Therefore, shop floor issues, such as bottlenecks, set-ups, lot sizes, priorities, random fluctuations and performance measurements, are treated in great depth. OPT maintains that the conventional assumptions made about the operation of the shop floor are mainly responsible for the poor performance of manufacturing in the past. The OPT philosophy incorporates ten rules which, when followed, are claimed to help move the organization towards the goal of making money. Eight of these rules relate to the development of correct schedules, while the other two are necessary for preventing traditional performance measurement procedures from interfering with the execution of these schedules.

The following sub-sections deal with each of these rules separately. The simple example which demonstrated the operation of MRP in Chapters 5 and 6, will be used here to highlight how some of the OPT rules can improve manufacturing performance. The example is set up to facilitate the explanation of OPT and does not represent the complexity that might appear in a true OPT installation.

14.3.1 Bottlenecks

A manufacturing organization can be considered as a system which transforms raw material into finished goods through the use of manufacturing resources. This general description of an organization could be applied to any facility, whether the finished goods were cars, oil, computers, etc. The manufacturing resources are crucial aspects in this process and can be considered to be anything that is required to produce the final product, whether it is a machine, an operator, space or fixtures. All of the resources in a manufacturing facility can be classified into **bottleneck** and **non-bottleneck** resources. A bottleneck can be defined as:

> '. . . a point or storage in the manufacturing process that holds down the amount of product that a factory can produce. It is where the flow of materials being worked on, narrows to a thin stream'. (Bylinsky 1983).

A bottleneck could be a machine whose capacity limits the throughput of the whole production process. Similarly, a highly skilled or specialized operator and/or scarce tools may be considered as bottlenecks.

Using the example from Chapter 7 where there were three work centres – assembly, painting and inspection – it can be shown what is meant by a bottleneck. If we recall, we have two different kinds of stool – Stool A and Stool B. From Table 7.6 we see that the processing times per unit for each work centre are as shown in Table 14.1.

Let us assume that each work centre can work for 40 hours per week and that, in the first instance, we are only producing one type of stool. Therefore, a market demand of only 100 units in a particular week leads to the following requirements for each work centre, as outlined in Table 14.2.

Table 14.1 Process times for each operation.

Assembly processing time	= 0.25 hours
Painting processing time	= 0.35 hours
Inspection processing time	= 0.20 hours

Table 14.2 Processing requirements for each work centre.

Assembly hours required	= 25 hours
Painting hours required	= 35 hours
Inspection hours required	= 20 hours

Table 14.3 Processing requirements for 130 units.

Assembly hours required	= 32.5 hours
Painting hours required	= 46.5 hours
Inspection hours required	= 26.0 hours

If the market demand were for 130 units the work centre requirements would be as shown in Table 14.3.

Assuming a 40 hour week, this clearly shows that the painting work centre can be considered a bottleneck when the demand is for 130 units. With respect to the painting work centre, the other work centres are considered as non-bottlenecks.

When we consider the manufacturing resources, some of which are bottlenecks and some of which are non-bottlenecks, certain relationships exist between the resources. OPT highlights four basic relationships that can exist between a bottleneck and a non-bottleneck. These are illustrated in Figure 14.2 (adapted from Fox (1982a)).

Each of these four relationships will now be discussed in turn.

- **Type I relationship** In this relationship, all products flow from the bottleneck to the non-bottleneck resource. This relationship occurs in the example above. Here, the painting work centre may be fully utilized at 100%, but the inspection work centre can only be utilized for

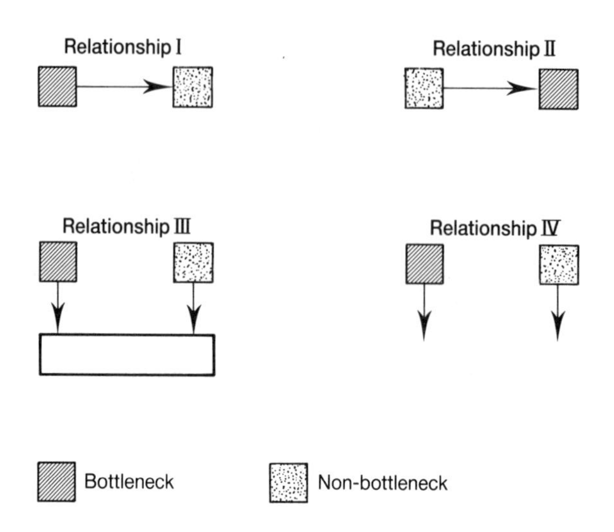

Figure 14.2 Relationships between bottlenecks and non-bottlenecks.

approximately 80% of the time. This is due to the painting work centre *starving* the inspection work centre of raw material. If the demand increases beyond this point, the inspection work centre, in theory, is able to meet the increased work load, but it cannot because of the metering effect of the painting work centre in restricting the flow of material.

- **Type II relationship** In this situation, all products flow from a non-bottleneck to a bottleneck. This again occurs in the example above, where the assembly work centre feeds the painting work centre. As shown above, if the assembly centre is utilized approximately 71% of the time, the painting work centre is already utilized 100% of the time. However, the utilization of the assembly work centre could be increased to 100% provided the market demand required it, but this would lead to a continuous build-up of inventory at the painting work centre. This leads to an increase in inventory without any change in throughput for the overall system.
- **Type III relationship** The third relationship is where a bottleneck work centre (let us call this work centre **leg-form**) and a non-bottleneck work centre (let us call this work centre **seat-form**) feed a common assembly work centre. Let us assume that the assembly work centre requires one unit from each work centre to complete a product. If work centre leg-form (the bottleneck) is operating at 100% utilization, we require work centre seat-form (the non-bottleneck) to operate at 80% utilization to keep the assembly work centre supplied with material. If,

however, the demand increases we can increase the throughput of the non-bottleneck (work centre seat-form) but this will lead to a build-up of inventory before the assembly work centre since the bottleneck (work centre leg-form) prevents an increase in throughput (in the OPT sense).

- **Type IV relationship** The final relationship exists when two work centres feed independent market demands. Again, work centre leg-form (the bottleneck) is utilized 100% of the time, but unless the market demand for the products of work centre seat-form increases, it (the non-bottleneck work centre) is still limited to processing just 80% of the time.

In all four of the above relationships the same result was obtained, i.e. the non-bottleneck should work at a reduced level of utilization sufficient to support the bottleneck while, at the same time, preventing a build-up of WIP (Work in Progress) or inventory at the bottleneck station, and the bottleneck should work at 100% utilization. In reality, a manufacturing plant can be simulated by a combination of any of the above relationships. Thus, resources in a plant could either be labelled as bottlenecks or non-bottlenecks. The strategy suggested is to ensure that the bottleneck resources are fully utilized at all times. With regard to non-bottleneck resources, not all of their time can be used effectively and some of their time is therefore considered as enforced idle time. It is important to remember that in the OPT approach this *idle time* is not considered detrimental to the efficiency of the organization. If it were utilized, it would possibly result in increased inventory without a corresponding increase in throughput for the plant.

The first rule of OPT derives from this discussion and is as follows.

RULE 1

The level of utilization of a non-bottleneck is determined not by its own potential, but by some other constraint in the system.

As shown above in relationships I, II and III, the utilization of the non-bottleneck was determined by the bottleneck and not by its own capacity. In the fourth case, the constraint applied to the non-bottleneck is the market demand.

From this discussion it is clear that non-bottleneck resources should *not* be utilized to 100% of their capacity. Rather, they should be scheduled and operated based on other constraints in the system. If this were done, the non-bottleneck resources would not produce more than the bottlenecks can absorb, thereby preventing an increase in inventory and operating expenses.

OPT deduces another rule from this discussion concerning utilization of resources:

RULE 2

Utilization and activation of a resource are not synonymous.

Traditionally, utilization and activation were considered to be the same. However, in OPT thinking, there is an important distinction to be made between doing the required work (what we should do – activation) and performing work not needed at a particular time (what we can do – utilization). Thus, it is vitally important to schedule all non-bottleneck resources within the manufacturing system based on the constraints of the system, which are usually the bottlenecks.

For example, one can operate a non-bottleneck resource at 100% utilization. However, if only 60% of the output of this non-bottleneck resource can be absorbed by the following resource, which for the purposes of this example is assumed to be a bottleneck, then 40% of the utilization of the non-bottleneck is simply concerned with building up inventory. From the point of view of the non-bottleneck resource, we could argue that we are achieving 100% efficiency. From a system point of view we are only 60% effective. We are confusing the utilization of the non-bottleneck with activation. Utilization is concerned with **efficiency**. Activation is concerned with **effectiveness**. Thinking in terms of the discussion on mechanistic and holistic approaches to problems in Part I, we could argue that, in this context, efficiency is a reductionist criterion whereas effectiveness represents a systems measure of performance.

14.3.2 Set-up times

Let us now consider another important aspect of shop floor activities, namely set-up times. The available time at any resource is split between processing time and set-up time. This is illustrated in Figure 14.3. However, there is a difference between the set-up times on a bottleneck and those of a non-bottleneck.

If we can save an hour of set-up time on a bottleneck resource, we gain an hour of processing time. Relating this to the fact that bottlenecks are a limiting constraint on other resources (and the system as a whole), an hour of production gained at a bottleneck has far-reaching implications. It can be equivalent to an increased hour of production and throughput for the total system.

At a non-bottleneck resource, we have three elements of time, namely processing, set-up and idle time. Clearly, if we can save an hour of set-up time, we gain an hour of idle time since the bottlenecks still constrain the capability of the non-bottleneck. Consequently, an hour saved at a non-bottleneck is likely to be of no real value. There is, however, one advantage to reducing set-up times at non-bottlenecks machines. Due to a lower set-up time, more set-ups can be used and the batch or lot sizes can be reduced.

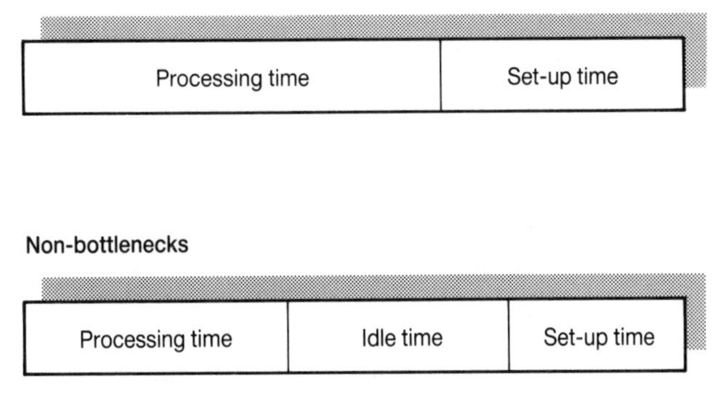

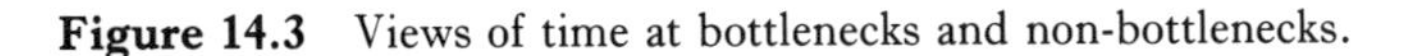

Figure 14.3 Views of time at bottlenecks and non-bottlenecks.

While a smaller lot size of itself does not increase throughput, it tends to reduce inventory levels and some operating expenses.

However, let us qualify this statement. For a specific sequence of operations to produce a specific product, a resource may be considered as a non-bottleneck. Moreover, for another sequence of operations and a different product, this same resource may be a bottleneck. Therefore, within the context of a specific schedule, saving an hour of set-up time at a non-bottleneck is *worth nothing*. On the other hand, within the context of all possible schedules, saving an hour at a resource which is a non-bottleneck in some schedules and a bottleneck in others, is equivalent to gaining an hour's production for the whole system. The next three OPT rules are deduced from the above points.

RULE 3

An hour lost at a bottleneck is an hour lost for the total system.

Therefore, to maximize system wide output, 100% utilization of all bottleneck resources should be a major goal of manufacturing.

RULE 4

An hour saved at an non-bottleneck is just a mirage.

Saving time at a non-bottleneck resource does not affect the capacity of the system, since system capacity is limited by the bottleneck resources.

OPT also maintains that the batch sizes for bottleneck resources should be as large as possible. This is in line with the thinking of the economic order quantity and economic batch quantity approaches (see Chapter 9). The reasoning behind this statement is that if we have many set-ups on a bottleneck resource, the amount of time it takes to do these set-ups is non-productive.

Therefore, we try to reduce the amount of set-ups, i.e. make the batch sizes as large as possible, in order to maximize the amount of productive time at a bottleneck. However, for non-bottleneck resources, there is no significant advantage in having large batches, in fact, the smaller the better as suggested above. The obvious solution to this is to use variable batch sizes. This represents a major departure from the materials requirements planning approach where a lot sizing technique normally determines a lot size that is used for all stages of production and transportation, regardless of whether these stages represent bottlenecks or not.

RULE 5

Bottlenecks govern both throughput and inventory in the system.

Traditionally, bottlenecks were believed to limit throughput only temporarily and to have little impact on inventories. OPT argues that inventories (particularly WIP) are a function of the amount of work required to keep the bottlenecks busy.

14.3.3 Lot sizes

Another important variable to be controlled on the shop floor is lot size. This is a crucial variable that is closely linked to the inventory and throughput of the organization. Traditionally, one lot size was determined as being optimal for the manufacturing process. OPT thinking differs from this and maintains that there should be two lot sizes. To demonstrate this let us take an example.

If we examine an assembly line to determine the lot size used, there are two possible answers. On the one hand, the lot size is frequently thought of as one, where one item is moved from one assembly station to another. Alternatively, the lot size can be considered to be infinite since the products on an assembly line are very infrequently changed. Both views are correct, depending on the perspective of the viewer. When we say that the lot size is one, we are viewing the process from the standpoint of the item or product in production. If we say that the lot size is infinite, we are viewing it from the standpoint of the resource.

An analogy may be helpful. Consider a road that is so hilly and convoluted that at no time can a traveller see any other travellers either in front or behind him/her. Let us also assume that all travellers on this road are moving at the same speed. (This represents the basic operation of an assembly line.) Then, from the perspective of a particular traveller, it would seem that he/she is the only traveller on the road. This is similar to the lot size seen from the product perspective. If someone were to sit by the side of the road and watch the travellers go by, to him/her the number of travellers passing would seem infinite. This is analogous to the lot size as seen from the standpoint of the resource.

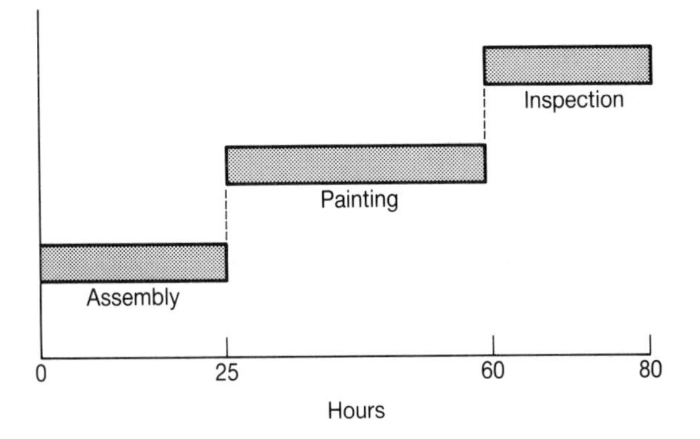

Figure 14.4 Process batch = transfer batch.

Therefore, from the OPT perspective, there are at least two lot sizes to be considered in manufacturing:

(1) The transfer batch – the lot size from the parts point of view.
(2) The process batch – the lot size from the resource point of view.

The next OPT rule is derived from this distinction.

RULE 6

The transfer batch may not, and many times should not, be equal to the process batch.

Lot splitting and overlapping of batches were traditionally discouraged in manufacturing. OPT maintains that the manner in which batches are processed is essential to the **effective** operation of a production system. Returning again to the simple example of Chapter 7, let us assume that we have to produce 100 units of Stool A. This will require processing time on the various manufacturing resources as outlined in Table 14.1. If the transfer and process batch are equal, then the batch of 100 units will not be moved to the painting work centre until all the stools have been assembled. This is shown in Figure 14.4.

Figure 14.4 shows that, using this procedure, it will take two weeks for the order to be processed. If, on the other hand, the transfer batch is not equal to the process batch, then it is possible to move a part immediately to the following work centre (as specified in the manufacturing routing information) once it has finished its present operation. Figure 14.5 illustrates the effect of this procedure.

The figures are not to scale but they do serve to show that there is a marked reduction in throughput time – something of the order of 40%. This,

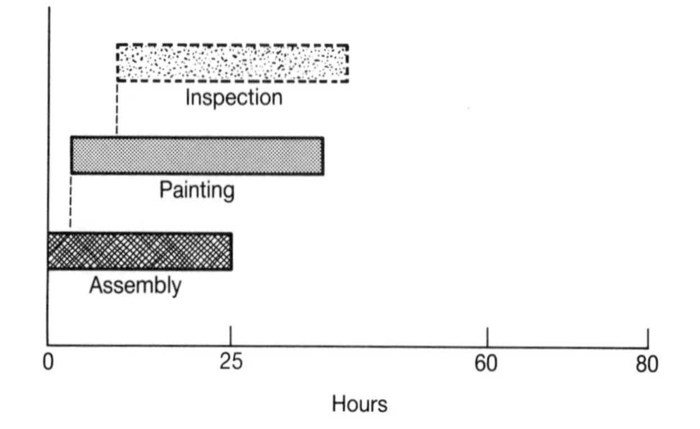

Figure 14.5 Process batch ≠ transfer batch.

of course, also reduces inventory and operating expenses. The dashed line showing the operation of the inspection work centre in Figure 14.5 represents the fact that the painting work centre is a bottleneck. As such, it can not produce enough parts to maintain full utilization of the inspection work centre.

Therefore, it is argued, it is impossible to determine from the outset a single lot size that is correct for all operations. This concept is encapsulated in another OPT rule.

RULE 7

The process batch should be variable, not fixed.

This implies that the process batch size at different work centres should not be the same. Traditional manufacturing practice would suggest that, except in exceptional cases, the batch size should be fixed, both over time and from operation to operation. However, in the OPT approach, process batches are a function of the schedule and potentially vary by operation and over time.

The lot size is established dynamically for each operation and balances inventory cost, set-up costs, component flow requirements and the needs for managerial control and flexibility. In addition, some operations might be bottlenecks and may require large process batches, while non-bottlenecks may require small process batches in order to reduce lead time and the resulting inventory.

14.3.4 Lead times and priorities

MRP, reviewed in Part II, is based on an assumption that planning lead times can be determined *a priori*. As seen in Chapter 5, lead times are used to

Table 14.4 Calculation of production time.

Production lead time for Stool A		
Batch size	× Processing time per unit	
100	× 0.35 hours	= 35 hours

Production lead time for Stool B		
Batch size	× Processing time per unit	
100	× 0.35 hours	= 35 hours

offset from the identified due date in order to calculate the time to start production or to release the purchase orders.

Thus, MRP uses the estimated lead time to determine the order in which jobs are processed. Priorities are assigned to jobs and those with the higher priorities are processed first. The estimated lead time is, in turn, dependent on the estimated queuing time for each operation. Once priorities have been established, the capacity of the production process is examined to see if the plan can be met. However, the important interaction between priority and capacity is not examined. Priority and capacity are essentially considered sequentially, not simultaneously.

The example of the two stools and Gizmo-Stools Inc. will be used to clarify this point. Consider the painting work centre and assume that we have two orders, one for a quantity of 100 of Stool A and a second for a quantity of 100 of Stool B. If, for argument's sake, we further assume that we have two painting work centres, then the production lead time for both batches would be as shown in Table 14.4. (Note that for the purposes of this discussion, set-up time is ignored.)

In this case, we can say that the production lead time for this work centre is 35 hours. However, if, in fact, we have only one painting work centre then the following applies.

If Stool A is processed first, the production lead time for Stool A is 35 hours and Stool B must wait for these 35 hours for the painting work centre to become available. Thus, the overall production lead time for Stool B is, in effect, 70 hours. If Stool B is processed first, the reverse applies. Thus, if we schedule in one fashion we get one set of lead times and if we schedule another we get another set of lead times.

This illustration demonstrates that:

- Actual lead times are not fixed.
- Lead times are not known *a priori*, but depend on the sequencing at the limited capacity or bottleneck resources. Exact lead times, and hence

priorities, cannot be determined in a capacity bound situation unless capacity is considered.

From this example and the points it highlights, another OPT rule is deduced.

RULE 8

Capacity and priority should be considered simultaneously, not sequentially.

So far, eight rules have been described and the thinking behind them explained. These rules focus primarily on the operation level and particularly the scheduling of work – through the shop floor. The next two OPT rules are concerned with the performance measures used to monitor the effectiveness of the shop floor.

14.3.5 Cost accounting and performance evaluation

The rules so far are related to the development of *correct* schedules. It is important that once these schedules are developed, they are followed in detail. The OPT philosophy identifies a number of barriers to the implementation of these *correct* schedules. Some of the major obstacles are briefly described below. According to the proponents of OPT these are:

- Methods of measuring efficiency.
- The expectation of balanced plant loads.
- The so called **hockey stick** phenomenon.

Each of these will now be discussed in turn.

Measures of efficiency

According to OPT thinking, one of the greatest threats to the use of good schedules is the *misuse* of cost accounting procedures in performance measurement systems. Cost accounting principles, when used to measure performance, are in conflict with OPT rules 3 and 4 which state that *an hour lost at a bottleneck is an hour lost for the total system* and *an hour saved at a non-bottleneck is just a mirage*. Present day cost accounting practice does not differentiate between work at a bottleneck and work at a non-bottleneck resource. To illustrate the consequences of this, consider the situation of a non-bottleneck resource feeding a bottleneck resource in the plant.

In traditional manufacturing management practice, the supervisors of both types of resources are encouraged to seek 100% resource efficiency since, in general, they are measured by their production rates and not by how well their output impacts the output of the total manufacturing organization. If the supervisor of the non-bottleneck resource operates at full capacity this will result in a costly build-up of inventory and increased operating expenses due to the inability of the bottleneck resource to absorb the production of the non-bottleneck resource. The bottleneck supervisor should always be operating his/her resources at full capacity. The performance of the supervisor who is dealing with non-bottleneck machines should be measured, not by the amount of WIP (work in progress) he/she is responsible for creating, rather by the volume of usable product that his/her area produced.

The supervisors are also encouraged to reduce the number of set-ups on all machines, whether bottleneck or non-bottleneck, in order to increase the *efficiency* of the machines. If supervisors were encouraged to reduce the lot sizes, i.e. increase the number of set-ups on non-bottleneck machines to feed a bottleneck machine correctly, this might well result in a better flow of product to the bottleneck machines, thus ensuring that they were never starved of work (remember, an hour lost at a bottleneck resource cannot be recovered and leads to an equivalent fall in system output) as well as lower inventory levels.

The two preceding paragraphs highlight how management emphasis on certain performance measures can, in some situations, lead to overall system inefficiency. Existing conventional measures of machine and operator performance tend not to take a systems approach, that would consider the interrelationships between all of the elements of the manufacturing system.

In effect, OPT argues, cost accounting principles attempt to measure *efficiency* of resources, not their *effectiveness*. OPT argues that it is important to understand that, from the perspective of the whole manufacturing organization, it is the *effectiveness* of each resource that is important.

The expectation of balanced plant loads

Manufacturing management, certainly in the Western world, has traditionally tried to manage the operation of the production system by the control of capacity of that system. If the utilization of a given resource is considered too low or too high, schedules are changed to counteract this and to balance the load across the whole plant. This sometimes results in virtually full use of all resources, without consideration of the relationships between resources and the effect of these utilization levels on the operation of the manufacturing organization as a whole. Some of the pitfalls of this approach were pointed out in the previous section.

An alternative approach to controlling the operation of production is to consider those products that are sold, or will be sold, to customers. A manufacturing plant should produce only what is ordered by customers or what can reasonably be expected to be ordered by customers. Each work centre should only produce what is required at the next work centre, and so on, until the plant only produces what is required overall by customers. This reduces costly inventory, saves on operating expenses and facilitates maximum throughput. Therefore, it is argued, one should attempt to balance the flow of products through the plant rather than the plant capacity. There is an OPT rule to this effect.

RULE 9

Balance flow not capacity.

Traditionally, the approach was to balance capacity and then to attempt a continuous flow. Line balancing is a good example of this approach. The work involved in manufacturing a product is divided into roughly equivalent elements from the capacity point of view. The resources involved in the production process are examined and their capacities balanced in order to ensure high utilization factors. Production then involves trying to create a continuous flow of material through these resources. OPT argues *against* balancing capacity and *for* a balancing of flow within the plant (similar to JIT). This involves looking at the product and ensuring that there is a continuous flow of material during production. The emphasis is on the flow of the product rather than on the resources used to produce it. This leads to the identification of bottlenecks, which can be then be examined with a view towards an increase in their throughput and, consequently, the throughput of the total system.

It is important to realize that the OPT approach and the JIT approach do not advocate disregarding capacity considerations. Rather, they suggest that production is controlled by considering product flow and capacity considerations simultaneously, not sequentially, as is the case in the MRP approach.

The hockey stick phenomenon

The developers of OPT identify what they term the *hockey stick* phenomenon and argue that it is caused by the conflict between two measurement systems – cost accounting and financial performance and is visible in most plants at the end of each financial reporting period. The hockey stick phenomenon is illustrated in Figure 14.6 (adapted from Fox (1982b)).

At the beginning of each period the plant is driven by cost accounting performance measurements, which have a *local* focus as discussed earlier. Measurements focus on machine and operator efficiency, standard times or costs to produce a part at a particular operation. To be *efficient*, large batch sizes are run through operations, regardless of whether they are bottleneck or non-bottleneck, usually resulting in the build-up of unnecessary inventory.

As the end of the financial reporting period approaches, management becomes concerned with a global measurement – the performance of the total system. There is an enormous effort to ship products, to make more money.

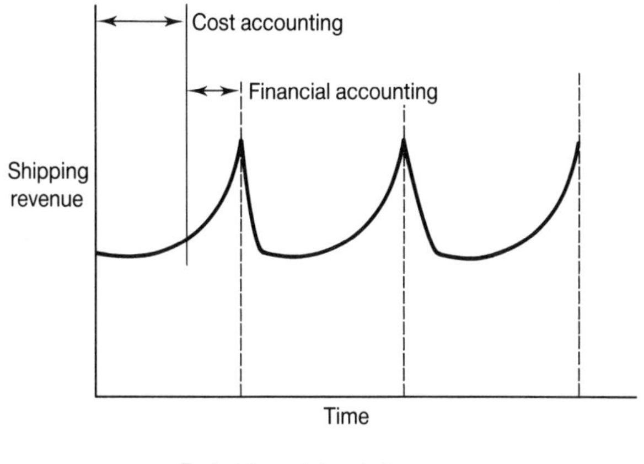

Figure 14.6 Cost accounting and financial measures.

Efficiencies, the number of set-ups etc., are no longer a consideration and, in their stead, expedited lots are split, overtime is allowed, inefficient machines are put working again, anything to increase shipments! Once this end of period is over, *local* efficiency measures take over again and the cycle repeats itself.

This method contrasts sharply with the OPT approach, which seeks to measure the performance of the plant as a whole on the basis of its raw material input and final product output, rather than by measuring only the efficiency of individual operators or machines or other elements of the subsystem. The final rule of OPT is a reflection of this thinking.

RULE 10

The sum of local optima is not equal to the optimum of the whole.

14.4 Conclusion

The preceding sections have covered the thinking behind OPT. Ten rules effectively articulate this thinking and these have been described. Effective operation of the manufacturing plant depends on the production of realistic and correct schedules.

Chapter 15 attempts to describe the software system developed by the promoters of OPT, which is based on the ideas behind the rules just described. Employing the rules outlined above will, it is argued, result in considerable improvement in manufacturing performance. However, introducing the OPT software gives one the ability to produce and follow *correct* schedules.

CHAPTER FIFTEEN

Optimized production technology system

15.1 Introduction

By simply applying the *ideas and rules* described in Chapter 14, 'companies can improve throughput, cut inventory and increase sales . . .' (Bylinsky (1983)). Combining the application of these ideas with the OPT (Optimized Production Technology) software system, even better results can be achieved, it is claimed. The OPT software system is based on a closely guarded algorithm which concentrates on identifying and scheduling the bottlenecks in the manufacturing system. Although the detailed operation of the software system is proprietary information, the basic operation of the software is understood. This chapter will outline how the system operates and then support this by developing OPT schedules for the example of Gizmo-Stools Inc.

15.2 The software system

In order to operate the OPT system, a complete description of the manufacturing system must be generated. This is accomplished in different stages by a module called *BUILDNET*. Firstly, a product network is constructed for each product manufactured within the organization, which shows exactly how it is manufactured and/or assembled, the resources required, etc. Figure 15.1 (Jacobs (1984)) represents an illustration of such a network. The product network contains information similar to that which would be maintained in the bill of materials and routing files for a conventional MRP system. (See Chapter 8.)

At the top of Figure 15.1 are the market requirements which are linked to the various assemblies or manufactured products. These products, in turn,

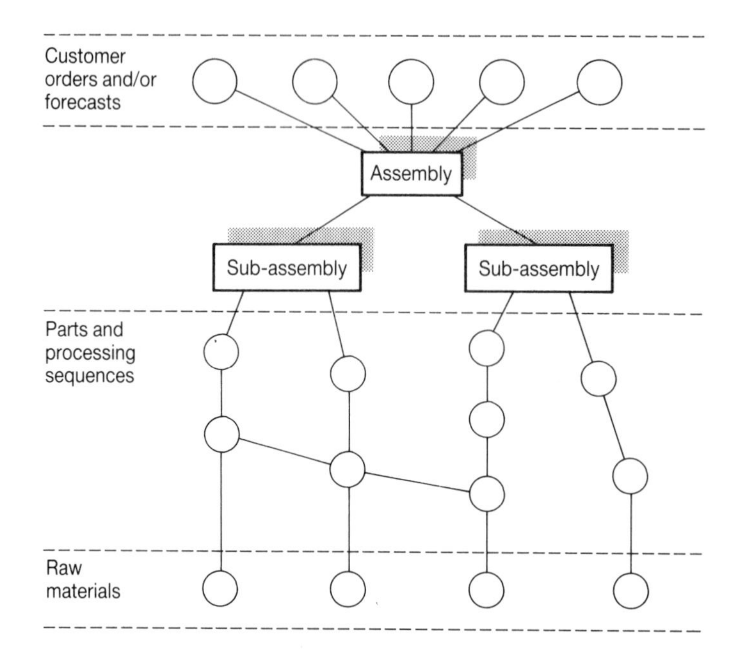

Figure 15.1 Sample product network.

are linked to the sub-assemblies from which they are assembled. The sub-assemblies are chained to the parts from which they are produced and the various manufacturing processes that these parts must undergo. Finally, the first manufacturing operation is tied in to the appropriate raw materials.

Detailed descriptions of each resource are then defined in OPT and combined with the product network to form an engineering network, as illustrated in Figure 15.2 (Jacobs (1984)). In an initial analysis, a module called *SERVE* uses this information to backward schedule to the order due dates specified in the product network. *SERVE* assumes infinite capacity is available at each resource and uses timing information calculated from the set-up time, run time and scheduled delay data included in the engineering network. The sole purpose of this analysis is the identification of critical bottleneck resources. A report showing utilization of each resource is developed. Those near or greater than 100% utilization are identified as bottlenecks that are important to the *SPLIT* operation performed next.

The *SPLIT* module separates the engineering network into two sections. As shown in Figure 15.3 (Jacobs (1984)), the upper section includes operations that use bottleneck resources and all operations that follow the *critical* operations. The lower section includes operations that precede the critical operations, i.e. all the non-bottleneck resources between raw material acquisition and the first bottleneck resource.

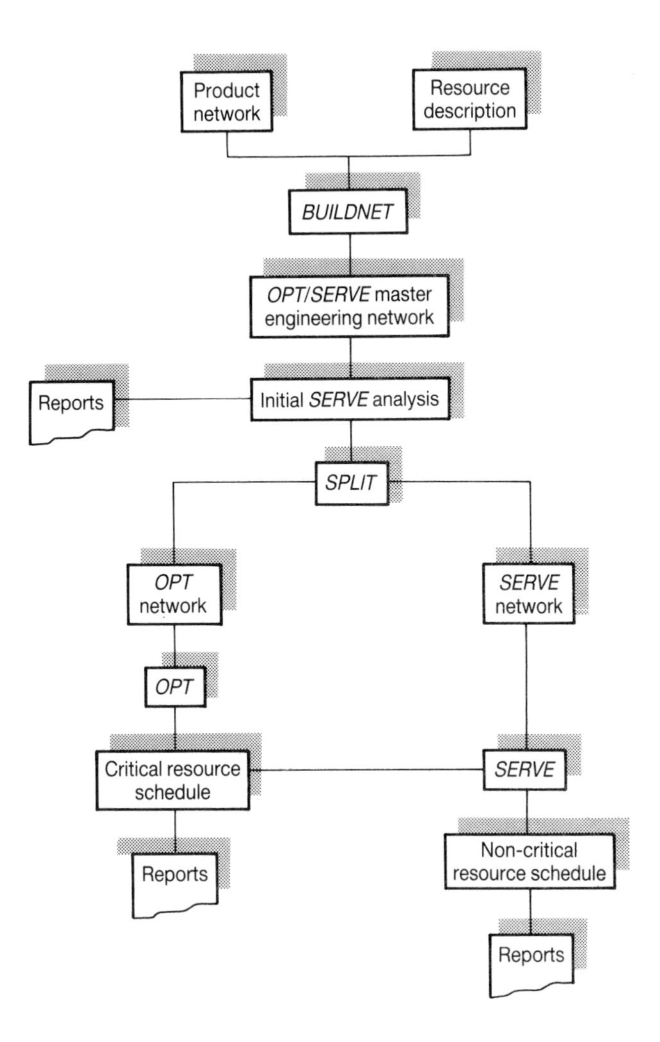

Figure 15.2 OPT/SERVE information flow.

The upper section of the engineering network is scheduled using the *OPT* module. The *OPT* forward scheduling procedure is based on a secret algorithm developed by Goldratt (1980) and takes into account the finite capacity of resources. This algorithm, as far as is generally known, is based on the rules described in Chapter 14 which deal with bottlenecks, set-ups, lot sizes, etc., to mention but a few. Once the critical part of the network has been scheduled, the non-critical or lower part of the network is scheduled using *SERVE*.

SERVE, as described earlier, is a backward scheduling procedure that assumes infinite capacity. In the initial use of *SERVE*, the due dates used are

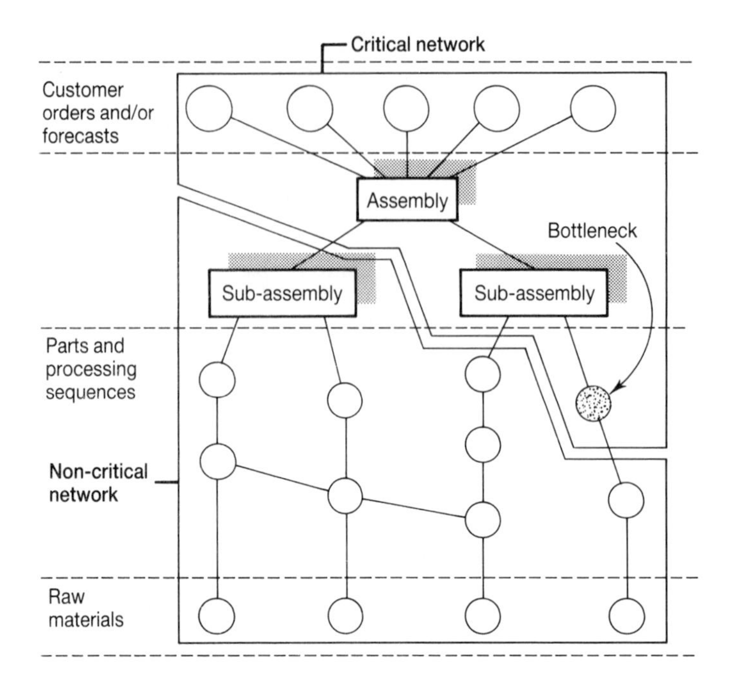

Figure 15.3 Product network – critical, non-critical split.

the order due dates. Now the due dates are those determined by the *OPT* module for the bottleneck resources in the critical section. In particular, *SERVE* schedules so that material will be available for the first operation in the critical part of the network. Referred to as the *OPT/SERVE* schedules, these involve the composite schedule, developed using the finite forward loading *OPT* module on the critical part of the network, followed by the infinite, backward loading of the non-critical part, as in Figure 15.2.

When all the OPT modules have come up with a complete production schedule, the schedule is run through the program, which may locate additional bottlenecks. These may have been created in the backward scheduling of the non-critical part of the network. These bottlenecks are then resubmitted to *OPT/SERVE*. After a number of runs through the system, an OPT schedule is complete.

It is claimed that there are management parameters within OPT which permit the fine tuning of the schedule to accomplish specific objectives. These are specified by the user. For example,

> 'if a company's goal is to be the lowest cost producer, OPT can be told to favour minimizing set-up costs at some expense to delivery performance. However, if delivery performance is more important, the adjustment can be made so that when there is a choice, OPT will favour delivery performance over set-up costs' (Fox (1982b)).

The above paragraphs have briefly outlined the operation of the OPT software in the production of schedules for the shop floor. This explanation of OPT is very simple and necessarily brief, largely because the authors do not have access to the source code of the OPT product. Section 15.3 will attempt to explain OPT using the Gizmo-Stools Inc. example that was used to explain the mechanics of MRP and of Kanban.

15.2.1 OPT as a productivity improvement tool

The developers of OPT claim that it can be used in numerous ways to improve productivity. It can be used not only as a production scheduling tool (as described above), but also as an analytical technique for 'simulating, analyzing and optimizing production operations' (Fox (1982b)). There are three major ways that management are using OPT to provide answers to *what if* questions.

(1) Changing the factory load (production requirements) and studying the impact on throughput, inventory and operating expenses, etc.
(2) Varying the manufacturing capacity by adding or removing resources.
(3) Modifying management policies to understand how these policies affect operating performance.

The OPT software, it is claimed, provides management with an ability to develop realistic schedules and, to a certain extent, answer *what if* questions. 'To date there has not been a comprehensive comparison of OPT schedules to those produced using conventional scheduling logic' (Jacobs (1984)). Likewise, OPT's ability to answer *what if* questions has not been determined, largely as a result of the relatively recent availability of the OPT product.

15.3 An illustrative example

To explore further the application of the OPT system we will attempt to create OPT schedules manually for the example of the Gizmo-Stools Inc. stool manufacturing activity, as described originally in Chapter 5. A word of caution is in order here. The example is set up to facilitate the explanation of the logic of OPT and does not seek to present complexity and the data structures which might appear in an actual OPT installation. Furthermore, the authors are not party to the Goldratt algorithm used within OPT and hence cannot develop a complex scheduling example. Nevertheless, the example should be useful in that it allows some appreciation of the OPT approach to shop floor scheduling.

As described in Chapter 5, it is assumed that Gizmo-Stools Inc. manufactures two types of stool, namely, a four legged stool and a three legged stool. The product structures are shown in Figure 15.4.

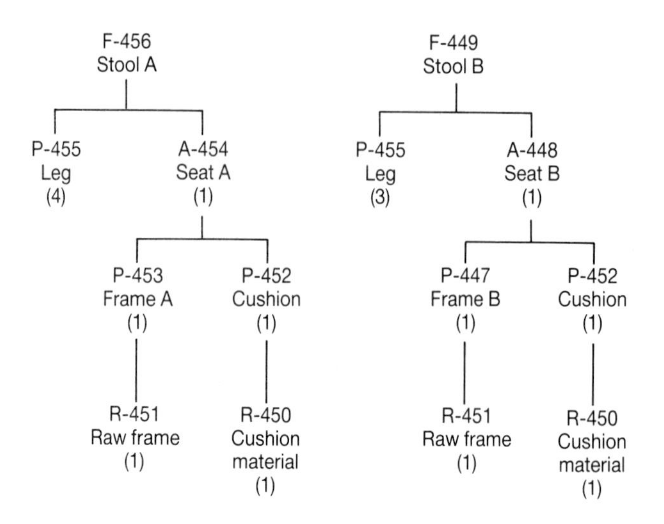

Figure 15.4 Two product structures.

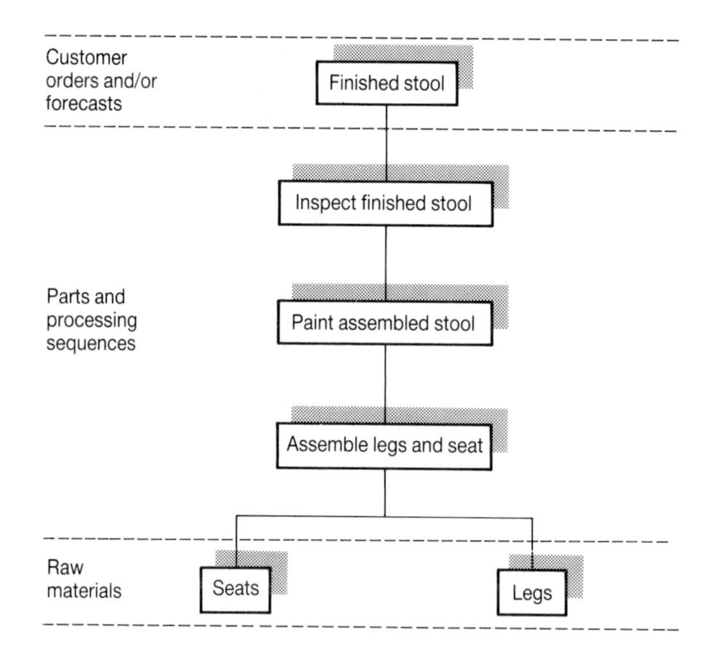

Figure 15.5 Stool product network.

From the information supplied above and the description of the process steps given in Chapter 7, the product network for the stool can be established as shown in Figure 15.5. Table 15.1 details the times for each operation and the resource which performs that operation for Stool B.

Table 15.1 Operation data to process Stool B.

Process requirements for Stool B
Number of operations 3 (All times in hours)

Operation number	*10*	*20*	*30*
Description	Assemble legs to stool	Paint stool	Inspect stool
Set-up time	0.5	0.75	0.5
Processing time	0.25	0.35	0.2
Operator time	0.25	0.35	0.2
Transport time	1.00	1.00	1.00
Work centre	Assembly	Painting	Inspect
Next operation	20	30	Stock room

Table 15.2 Processing time at each resource.

Operation number	**Resource**	**Time**
10	Assembly	25 hours
20	Painting	35 hours
30	Inspection	20 hours

Before proceeding with the example, two points are worth noting about Table 15.1 (which contains the information required by OPT and is similar to the information required by MRP as shown in Table 7.6):

(1) Unlike the MRP explosion process, fixed recommended batch sizes do not apply in the schedule generation process within OPT. Batch sizes are considered a function of the schedule and are determined later in the detailed scheduling process.

(2) Queue delay times are not used in OPT unless they are scheduled delays within the process itself, for example, time to allow paint to dry. In contrast, MRP includes queue times in the determination of lead times as seen earlier.

15.3.1 A simple example – an order for one product

To return to the example, let us assume that at the beginning of a week we have an order for 100 units of Stool B to be delivered at the end of the same week. For the purposes of demonstrating the scheduling activity of OPT, let us also assume that sufficient raw materials are available to meet this demand. Our problem is then one of scheduling the three resources. Neglecting set-up times for the moment, the total process times required at each resource to deliver the complete order are shown in Table 15.2.

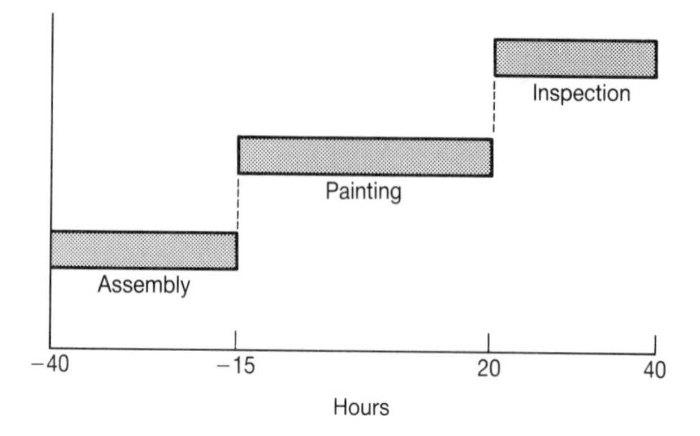

Figure 15.6 Standard MRP schedule.

If we use the standard MRP approach and, say, a batch size of 100, then in attempting to schedule this order we will produce a schedule as in Figure 15.6 which shows that we should have started this order one week ago if we are to meet the order deadline!

To meet this order using an MRP approach means *rushing* the order through the three work centres and, perhaps, interfering with other outstanding orders that are on schedule. This method of hurrying orders through the shop floor is termed *expediting* and is common in many companies. We have assumed a batch size of 100 in the above example, but there would still be a need for expediting even if the recommended batch size of 20, as used in the MRP example, were used.

If, instead, we use the OPT approach, then by backward scheduling from the order due date (see the description of the *SERVE* module above) and keeping in mind the difference between the transfer batch and the process batch, we produce the schedule given in Figure 15.7.

The important points to note about the schedule shown in Figure 15.7 are:

- The build-up of work in progress. The assembly work centre produces material which will queue before the painting work centre since the painting work centre is the bottleneck. Similarly, the output of the painting work centre initially builds up at the input to the inspection work centre.
- The scheduling of this order has assumed a transfer batch of one unit which may not be realistic for many industrial situations. For example, the transfer batch in the heat treatment of parts will generally be the capacity of the oven – or the pallet that is placed in the oven – since it may not be feasible to operate the oven for single parts. Conversely, on an assembly line, the transfer batch will rarely be greater than one.

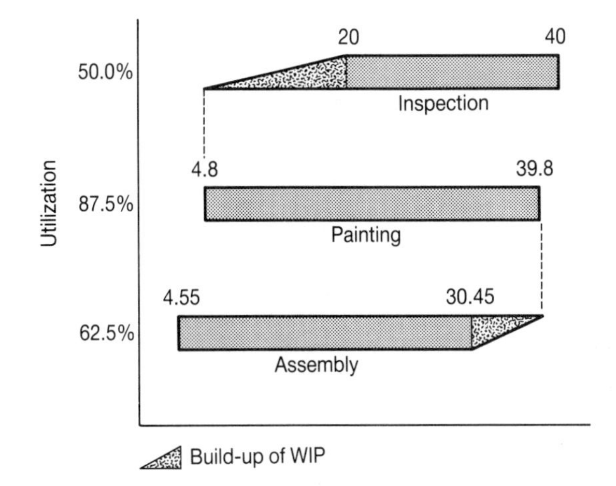

Figure 15.7 Initial OPT schedule.

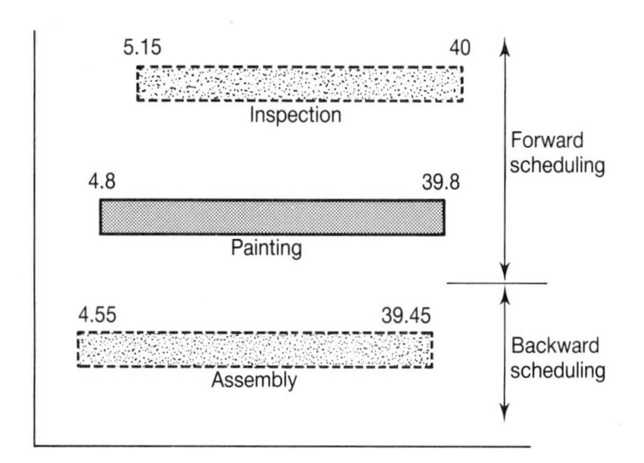

Figure 15.8 Final OPT schedule.

The main objective of producing the above schedule is to identify the bottlenecks and the utilization figures of each work centre show that the painting resource is the bottleneck. Once the bottleneck resource has been identified, all operations after and including the bottleneck operation are combined to give the *critical network*. All operations preceding the bottleneck are termed *non-critical*. In this instance, therefore, the critical network consists of the painting and inspection work centres. The assembly work centre is the non-critical network.

The critical network is now forward scheduled and the rules of Chapter 14 are incorporated into this process. These rules deal with utilization of resources, process batches and transfer batches. The non-critical network is backward scheduled in order to ensure that the bottleneck

Table 15.3 Operation data to process Stool A.

Process requirements for Stool A
Number of operations 3 (All times in hours)

Operation number	*10*	*20*	*30*
Description	Assemble legs to stool	Paint stool	Inspect stool
Set-up time	0.5	0.75	0.5
Processing time	0.39	0.35	0.2
Operator time	0.39	0.35	0.2
Transport time	1.00	1.00	1.00
Work centre	Assembly	Painting	Inspect
Next operation	20	30	Stock room

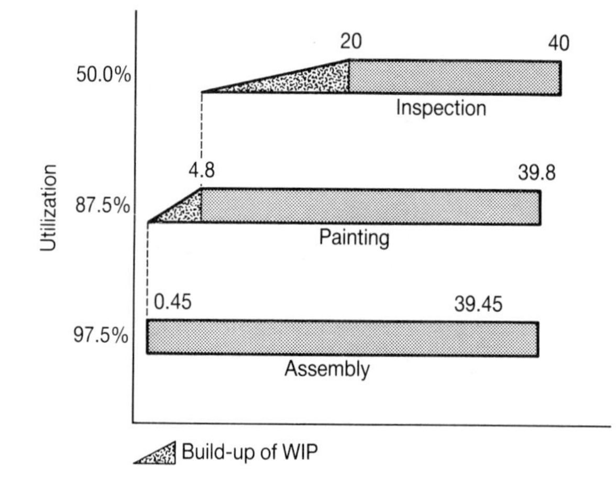

Figure 15.9 Initial OPT schedule for Stool A.

resource is not starved of material and is kept busy at all times. This results in the schedule depicted in Figure 15.8.

The dashed lines indicate that OPT schedules the inspection and assembly work centres so that they do not cause a major build-up of work in progress. Each non-bottleneck work centre is scheduled to be inactive for certain amounts of time.

The next step in the OPT system is to examine the non-critical section of the network to determine if any other bottlenecks have been created by the schedule generated for the critical section. Clearly, the utilization of the assembly work centre has not changed so the schedule, as given above, seems

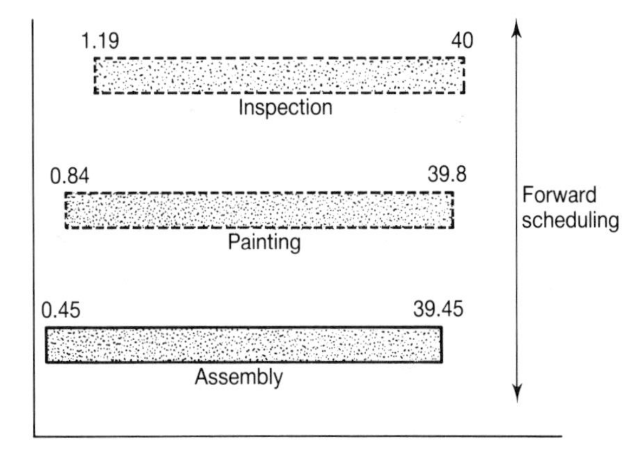

Figure 15.10 Final OPT schedule for Stool A.

final. It should be stressed again that this example does not present the complexity that would appear in a true OPT installation.

15.3.2 An order for two products

To illustrate how the level of complexity increases, we will consider a situation where there are orders for both types of stool. Before we do this, we shall briefly examine how an order for 100 units of Stool A would be scheduled using OPT. Table 15.3 contains the processing information for Stool A and the major difference is that the assembly of four legs to the stool, rather than three as in Stool B, takes longer. Our initial backward schedule is shown in Figure 15.9.

The points made above with respect to WIP and transfer batches also apply to this schedule. It is clear that the assembly work centre is the bottleneck, so that all three operations are included in the critical network. This critical network is forward scheduled to produce the schedule in Figure 15.10. The dashed lines again represent slack time, which results from the non-bottleneck work centres being scheduled in order to reduce WIP.

Returning to the situation of orders existing for both stools, let us assume that the order calls for 50 of Stool A and 50 of Stool B. Furthermore, we will assume that the order for Stool B takes priority over the order for Stool A, so that all B type stools must be produced on a work centre before type A stools can be started. Backward scheduling gives a schedule as illustrated in Figure 15.11.

From this we can see that the painting operation is again the bottleneck and causes the build-up of WIP. We therefore include the painting and inspection work centres in the critical network and forward schedule to produce a schedule as shown in Figure 15.12.

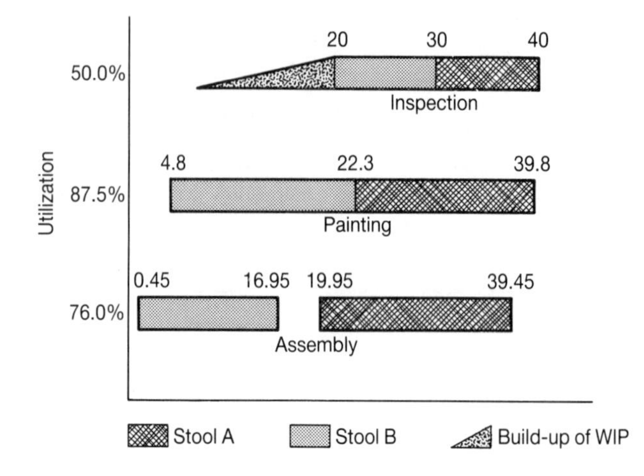

Figure 15.11 Initial two product OPT schedule.

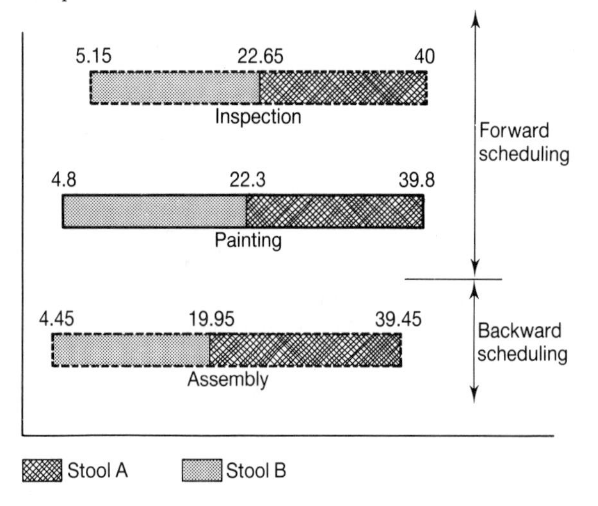

Figure 15.12 Final two product OPT schedule.

By examining the non-critical section, we see that we may have a bottleneck in the latter half of the week but, since there is enough raw material in stock, there is no need to reschedule.

Some observations about the above example will give an indication of the complexity of such an approach. We have assumed that all of Stool B should be produced before we start on Stool A. This is not a realistic assumption since, in many cases, a small batch of B may be followed by a small batch of A, resulting in many batches overall. This *mixed model* type

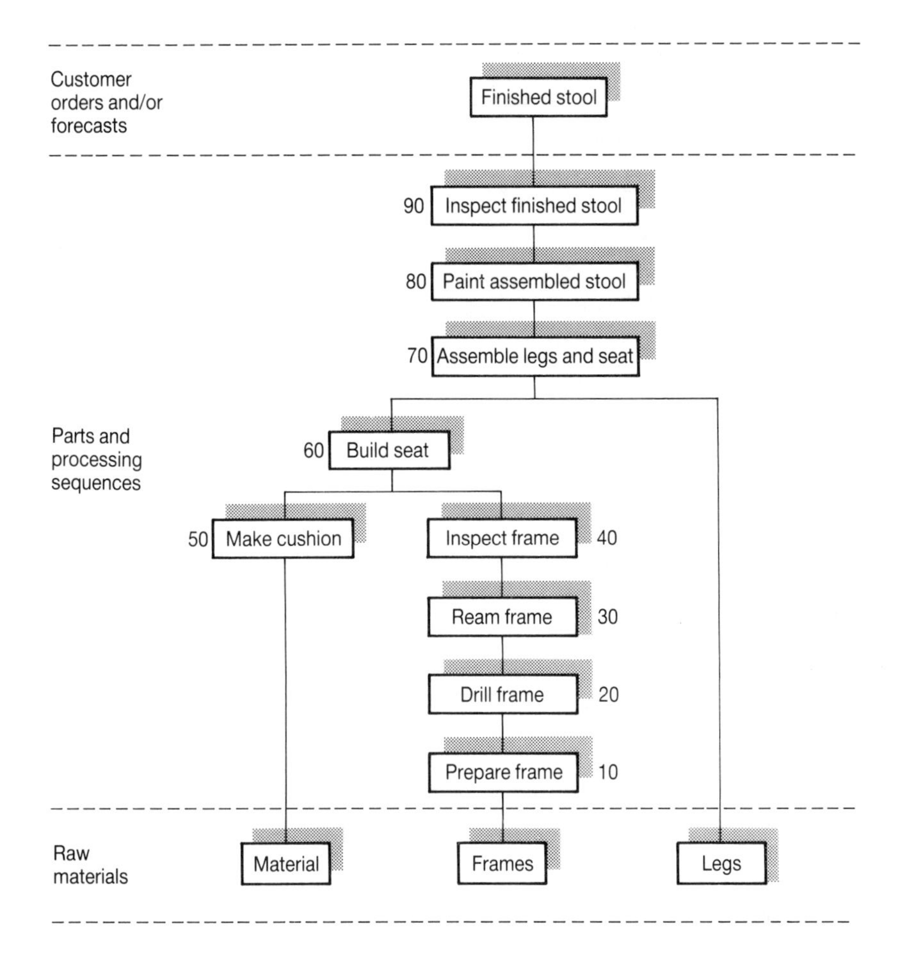

Figure 15.13 More specific product network.

production might work if the set-up times involved in changing from three legged stools to four legged stools, and vice versa, are relatively short. Here again, we have assumed a transfer batch size of one as above. This batch size could be variable, i.e. one in the case of Stool A (the more difficult stool to produce) but maybe two or more in the case of Stool B. This is in direct contrast to the MRP approach where lot sizes for MRP are generally equivalent to the order quantity. OPT clearly works on the basis of variable process and transfer batch sizes.

15.3.3 A further increase in complexity

The above example involves just two very simple products and three work centres. Each product follows the same sequence through the work centres

Table 15.4 Complete operation data to process Stool A.

Process requirements for Stool A
Number of operations **9** (All times in hours)

Operation number	*10*	*20*	*30*	*40*	*50*
Set-up time	0.5	0.3	0.4	N/A	0.5
Processing time	0.2	0.3	0.2	0.1	0.3
Operator time	0.2	0.3	0.2	0.1	0.3
Transport time	1.0	1.0	1.0	1.0	1.0
Work centre	Prep.	Drill	Ream	Insp1	Cushion
Next operation	20	30	40	50	60

Operation number	*60*	*70*	*80*	*90*
Set-up time	0.5	0.5	0.75	0.5
Processing time	0.1	0.2	0.35	0.2
Operator time	0.1	0.2	0.35	0.2
Transport time	1.0	1.0	1.0	1.0
Work centre	Build	Assembly	Painting	Insp2
Next operation	70	80	90	Stock room

and has similar processing times (apart from the assembly operation), and each of the work centres is only visited once by a product. Therefore, it is clear that this example is trivial. A more realistic example would involve many different operations with varying processing times and manufacturing routes through the shop floor.

Let us consider a slightly more complex example where the seats for the stools sold by Gizmo-Stools Inc. are manufactured by the company. The seats are assembled from a cushion and a frame which goes through four manufacturing process stages – it is prepared, drilled with three or four holes, reamed and inspected. The product network for this seat manufacturing process is shown in Figure 15.13 and the complete processing information for both types of stool is given in Tables 15.4 and 15.5.

Figure 15.14 represents the initial backward schedule designed to determine the bottleneck resources. As can be seen, three bottlenecks have been found. The unbroken lines represent the operations that are bottlenecks, whereas the broken lines are the non-bottlenecks, which should be scheduled to ensure that the bottlenecks are never starved of material. Now we will determine the process and transfer batch sizes for non-bottleneck and bottleneck resources. The process batch size will be as large as possible for a bottleneck and its transfer batch as small as possible.

This example is only for one product and it can be seen how complex the scheduling problem has become – it is both difficult and pointless to attempt

Table 15.5 Complete operation data to process Stool B.

Process requirements for Stool B
Number of operations 9 (All times in hours)

Operation number	*10*	*20*	*30*	*40*	*50*
Set-up time	0.5	0.3	0.4	N/A	0.5
Processing time	0.2	0.3	0.15	0.08	0.4
Operator time	0.2	0.3	0.15	0.08	0.4
Transport time	1.0	1.0	1.0	1.0	1.0
Work centre	Prep.	Drill	Ream	Insp1	Cushion
Next operation	20	30	40	50	60

Operation number	*60*	*70*	*80*	*90*
Set-up time	0.5	0.5	0.75	0.5
Processing time	0.15	0.25	0.35	0.2
Operator time	0.15	0.25	0.35	0.2
Transport time	1.0	1.0	1.0	1.0
Work centre	Build	Assembly	Painting	Insp2
Next operation	70	80	90	Stock room

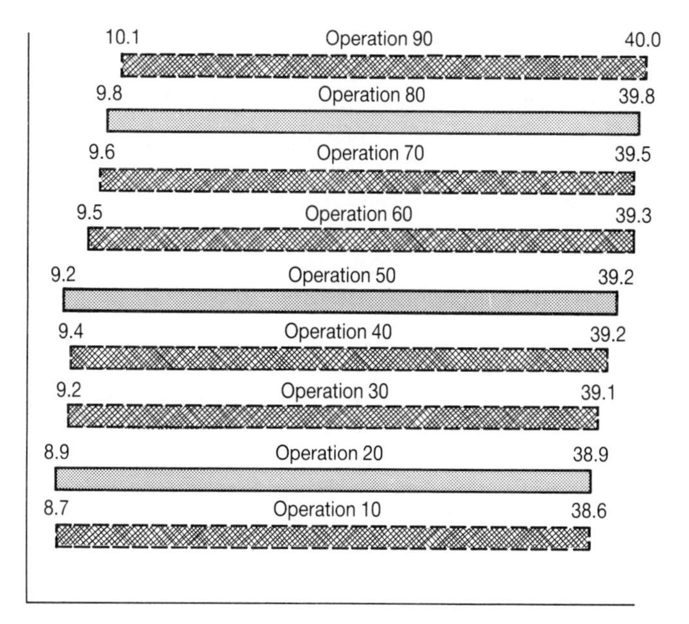

Figure 15.14 Initial schedule for the nine operations.

to develop the complete schedule. The need for software support to develop the full schedule is clear.

The above examples have shown our understanding of how the OPT software system might operate. The developers of the OPT product stress that the system will fail if the schedules it produces are not strictly followed. This is extremely important in OPT where safety stocks or other such buffers are not tolerated.

15.4 Requirements and assumptions of OPT

Clearly OPT requires a vast amount of data to develop the product network and the manufacturing model. Not only does OPT need to know how the product is made and through what processing route it passes, it also needs to have access to accurate set-up and run times for each individual manufacturing operation, maximum stock limits, minimum batch sizes, auxiliary machines, scheduled delays, etc. The user 'must already have a prodigious amount of precise data timing each step in the manufacturing process' (Bylinsky (1983)). However, many companies have already assembled such data for materials requirements planning systems. Existing data files, such as bill of materials, routings, inventories and work centre data are normally used by the OPT system.

When OPT was initially proposed, there seemed to be some suggestion that its data requirements were less rigorous than those of MRP type systems. The thinking was that one only needed accurate data on the critical or bottleneck resources and the products which visit those resources. This may not be the case. Firstly, the bottleneck resources may vary from time to time as the product mix in the shop load varies, thus one can never be sure what resource will become critical for however short a time. Secondly, given that OPT produces detailed shop schedules, it is vital that the data – particularly the process times on which this schedule is based – be accurate. If anything, in our view, the data requirements of OPT are more stringent than those of MRP.

An important aspect of the OPT approach is the need for shop floor supervisors and others (who are required to execute the schedule generated by the OPT software) to have confidence in the schedule presented to them by the computer. This confidence is clearly necessary since the schedule is expected to be followed rigorously. In our experience, this is an unusual approach to scheduling manufacturing operations and one which is at variance with the traditional freedom – indeed responsibility – of the shop floor supervisor to organize work within the area for which he/she is responsible. There have been reports of difficulties with some applications of OPT. Perhaps part of the difficulty stems from this aspect of the OPT approach.

15.5 Some views on OPT

OPT can be considered from a number of points of view. When OPT became available initially it was presented as a competitor for MRP (and MRP II) and JIT. In recent years there seems to be less emphasis on MRP versus OPT. Writers, both academic and practitioner alike, seem to be moving towards the view that the two are not incompatible.

It should also be said that when OPT was first made available, it attracted considerable criticism – and, indeed, continues to be criticized – because of the claim, implicit in its name, that it offers an optimal schedule and because of the fact that the scheduling algorithm on which it is based has never been revealed in the literature.

Lundrigan (1986) suggests that OPT brings together the best of JIT and MRP II into a 'kind of westernized just in time'. We agree that there is some truth in this statement, to the extent that OPT shares similar insights with JIT at the operational level, e.g. the use of small batches, the identification of transport and process batches, etc. However, OPT concerns itself with scheduling to the virtual exclusion of all else. JIT, as seen in Chapters 11 and 12, is concerned with establishing a manufacturing – indeed, a business – environment where shop floor control and, consequently, scheduling problems are minimized. We believe that JIT is much wider in scope than OPT, which concerns itself primarily with the generation of accurate shop floor schedules.

Swann (1986) argues that a company

> 'may, in fact, need both tools: MRP for net requirements and OPT for realistic shop schedules. Using OPT as a scheduling tool in, for instance, a job shop, does not preclude the need for accurate bills of material and disciplined inventory planning and control. MRP is the appropriate tool to provide bill of material and inventory management features'.

Vollmann (1986), coming from an academic background, offers a similar perspective and sees 'OPT as an enhancement to MRP II'. Vollmann argues that MRP II divides into three sections, 'the front end results in the master production schedule, the engine includes *little* MRP and capacity requirements planning and the back end completes the process – out to the shop floor and vendor follow up'. According to Vollmann, OPT can be used to evaluate an MPS from a capacity point of view, in order to determine its feasibility and OPT also outputs a 'detailed shop schedule that concentrates on the most important resources in the factory'.

We tend to agree with the points made by Vollmann, and that OPT is best considered as an enhancement to MRP II. After all, MRP II and OPT are similar in many respects. Each requires a large and complex production database. OPT requires that the process and product (i.e. bill of material) data be brought together to create the so-called product network, described earlier in this chapter. Much of basic data to achieve this is available from existing MRP II systems. The insights offered by OPT, if not the software itself, can be usefully applied in the shop floor or production activity control system.

15.6 Conclusion

This chapter has attempted to describe the relatively new production management system, OPT. Basically, OPT can be considered from two points of view – the OPT approach to manufacturing planning and control and the OPT software product.

The OPT approach to manufacturing planning and control is most often articulated in terms of ten relatively simple rules, as outlined in Chapter 14. Many of these rules represent an implicit criticism of traditional scheduling practice and of the metrics used to measure the performance of a manufacturing system. There is no doubt that the criticism of certain aspects of traditional manufacturing practice by the developers of OPT is valid – in particular, how cost accounting metrics can lead to system inefficiencies while *increasing* individual machine utilization levels. The insights concerning the relative importance of bottleneck and non-bottleneck resources, the use of separate process and transportation batches, etc. are very valuable to anyone concerned with the scheduling of work through a manufacturing system. These insights are offered in the form of OPT rules designed to reduce inventories and operating cost while simultaneously increasing the throughput of the manufacturing plant.

The OPT rules can, of course, be implemented without recourse to the second element of OPT described in Chapter 15, namely, the software package designed to produce realistic schedules. An important point about this software system is the requirement that the schedule it generates be followed exactly. This approach may cause difficulties in manufacturing systems where, traditionally, the shop floor supervisors considered a certain level of discretion with their operations schedules to be important.

Our understanding of OPT leads to the belief that it is best considered as an enhancement to the MRP II paradigm of production management systems.

References

Anonymous. 1984. 'Competitive analysis', *Automated Manufacturing Report*, (9) Frost and Sullivan.

Bylinsky, G. 1983. 'An Israeli shakes up US factories', *Fortune*, September 5th, 120–132.

Fox, R.E. 1982a. 'OPT: An answer for America. Part II', *Inventories and Production Magazine*, **2**(6).

Fox, R.E. 1982b. 'MRP, kanban or OPT: What's best?' *Inventories and Production Magazine*, July–August.

Fox, R.E. 1983a. 'OPT: An answer for America. Part IV. Leap-frogging the Japanese', *Inventories and Production Magazine*, **1**(2).

Fox, R.E. 1983b. 'OPT vs MRP: Thoughtware vs software', *Inventories and Production Magazine*, November–December.

Fox, R.E. 1983c. 'OPT: An answer for America. Part III', *Inventories and Production Magazine*, **3**(1).

Fox, R.E. 1984. 'Cost accounting measures of productivity: main bottleneck on the factory floor', *Management Review*, November, 55–58.

Goldratt, E. 1980. 'Optimized production timetables: a revolutionary program for industry', in *APICS 23rd Annual International Conference Proceedings*, 172–176.

Jacobs, F.R. 1983. 'The OPT scheduling system: a review of a new production scheduling system', *Production and Inventory Management*, **24**(3), 47–51.

Jacobs, F.R. 1984. 'OPT uncovered: many production planning and scheduling concepts can be applied with or without the software', *Industrial Engineering*, **16**(10).

Lundrigan, R. 1986. 'What is this thing called OPT?', *Production and Inventory Management*, **27**(2), 2–12.

Swann, D. 1986. 'Using MRP for optimized schedules (emulating OPT)', *Production and Inventory Management*, **27**(2), 30–37.

Vollmann, T.E. 1986. 'OPT as an enhancement to MRP II', *Production and Inventory Management*, **27**(2), 38–46.

PART V

PMS: a view of the future

Overview

In Part I of this book the context within which production management systems must operate was defined. The emergence of computer integrated manufacture was looked at together with the competitive pressures which push manufacturing firms to adapt integrated solutions. The key role of production management systems within CIM was pointed out and it was argued that PMS is at the very heart of an integrated manufacturing system.

In Parts II, III and IV the three approaches to production management systems were considered. The essential assumptions behind the three approaches were explored and their main features presented with a critique of each being offered. As was seen, MRP, JIT and OPT are not competing technologies. Each is different in terms of its scope, i.e. the range of problems it seeks to address, the types of manufacturing systems in which it has been applied and in terms of its ability to fit into an emerging *integration* environment.

In Part V, the main weaknesses of each approach will be discussed, as well as identifying the important insights which each offers. Although projecting the future is a dubious art, we shall proceed to develop an outline sketch of the attributes of an architecture for future hybrid production management systems. This, we believe, represents the most likely state towards which current systems will evolve.

Finally, with a view to operating the existing levels of production management technology as effectively as possible, a series of key insights will be offered, which we believe our study of JIT, MRP and OPT has uncovered.

CHAPTER SIXTEEN

Beyond MRP II, JIT and OPT

16.1 Introduction

This chapter aims to communicate an understanding of how ideas from the alternate production management paradigms – MRP II, JIT and OPT – can fit together. This problem is approached, not by attempting to force fit three competing approaches, but rather by searching out the fundamental weaknesses and strengths of each approach and then proposing a hybrid solution which addresses the requirements of PMS within CIM.

The discussion begins with a review of the scope and application areas of these technologies in the context of the new manufacturing environment and considers the difficulties of designing and implementing PMS within CIM. Next, the core theories and key failings of the three paradigms are examined, focusing on the essence of the approaches, as opposed to application issues.

An attempt is then made to take a very broad perspective on production management in a CIM environment and to position the available PMS solutions against this perspective. In many ways this picture plays the role of a murky crystal ball in which can be seen the future possibilities of the various PMS approaches (which seem to indicate the emergence of hybrid architectures). The difficulties in designing and installing PMS systems within a CIM environment are considered and an attempt is made to define a hybrid architecture in general terms. The final offering is a set of advice relating to production management systems in a CIM environment.

16.2 The scope of MRP, OPT and JIT

We will now try to establish a framework for articulating the scope of MRP, OPT and JIT. We do this by taking a simplified view of a manufacturing facility and neglecting, for a moment, considerations such as labour (both

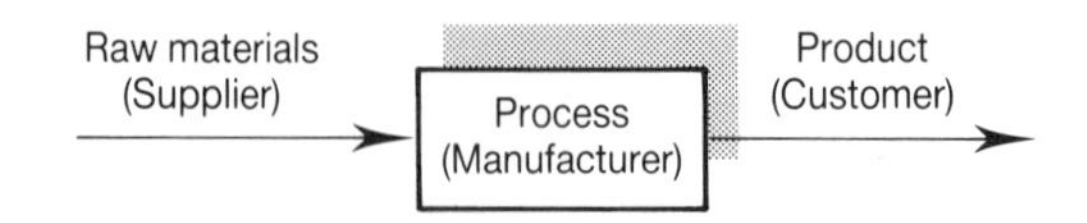

Figure 16.1 Simplified view of a manufacturing facility.

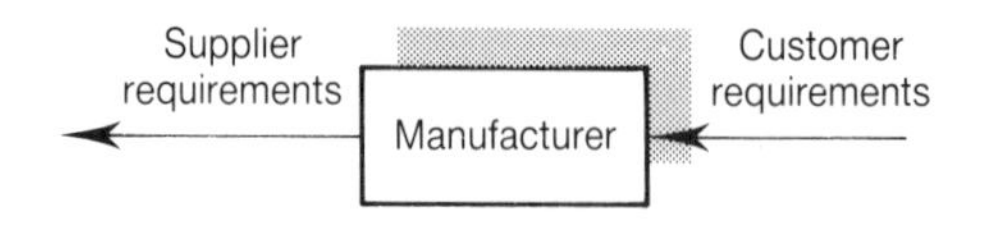

Figure 16.2 Simplified view of information flow.

direct and indirect), finance, etc. We will consider the manufacturing facility as a process which produces a product from raw materials as illustrated in Figure 16.1. This allows us to view the manufacturing facility from two perspectives – from the processing viewpoint or, alternately, from the product viewpoint. We can then examine how each of the three major production management systems sees the manufacturing facility from these two points of view.

If we enrich Figure 16.1 and include some of the information that passes between the three main actors – the customer who requires the finished product, the manufacturer who produces it and the vendor who supplies the manufacturer with the raw materials – we arrive at Figure 16.2.

From a management point of view it can be seen that it is mainly requirements that are passed between the actors. The MRP approach is concerned with the logistics of the manufacturing process. It takes the customer requirements for products and breaks these into time-phased requirements for sub-assemblies, components and raw materials. It also seeks to schedule activities within the manufacturing process and the availability of material from vendors to produce the product on time for the customer. It can be said that MRP is really only concerned with the logistics of *when*.

JIT takes a somewhat larger view and is concerned with *what* the product is, *how* the product is manufactured and the logistics of delivering it on time to the customer. JIT seeks to develop enduring relationships with vendors and, in so far as is possible, to *influence* the vendors' manufacturing process, in order to achieve JIT delivery of raw materials and purchased items.

OPT, in essence, is a scheduling tool and, like MRP, is mainly concerned with the logistics of *when* and *how many*. However, like JIT, it takes a more granular look at the production process and produces more detailed schedules for the shop floor. OPT also seeks to influence, to a limited extent, *how* the product is manufactured, through modified process and transfer batches, emphasis on bottleneck resources, etc.

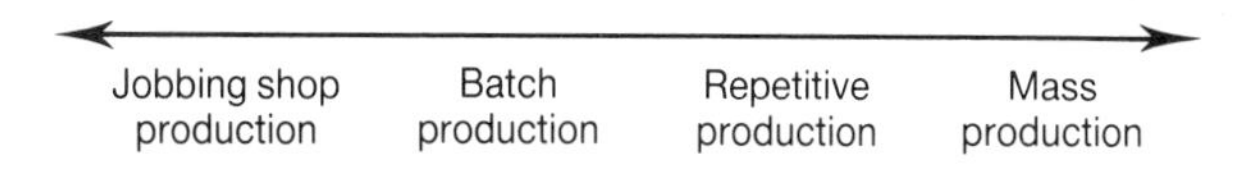

Figure 16.3 Continuum of manufacturing systems.

In terms of the scope of the three approaches, therefore, it may be argued that MRP and OPT are primarily concerned with *when*, whereas JIT seeks to influence the *what*, the *how* and the *when* of manufacturing.

16.3 MRP, OPT and JIT – application areas

Throughout this book, manufacturing systems have been considered to fall into three broad categories, namely, jobbing shops, batch production shops and mass production systems. In the discussion of JIT in Part IV, this classification was widened to include repetitive manufacturing systems.

In terms of the application of the various techniques, it is clear that people tend to associate JIT with mass production and repetitive manufacturing systems. JIT did, after all, originate in the final assembly plants of the major Japanese automotive manufacturing companies. In a similar vein, one might argue that MRP – and indeed OPT – thinking can be associated with batch production systems, given the fact that both systems tend to be concerned with a relatively large number of products, associated numbers of Bills of Materials (BOMs) and demand which is, at best, a combination of actual orders and forecasts and, at worst, a forecast.

Given the clear distinction between JIT philosophy, JIT techniques for manufacturing process design and planning, and the Kanban system, we would like to repeat the following observation. Kanban is essentially a Production Activity Control (PAC) system which only functions well in a mass production or repetitive manufacturing environment. However, the JIT philosophy and, indeed, JIT manufacturing and planning techniques, are applicable to all types of discrete parts manufacturing. Indeed, the greater the degree to which the JIT philosophy is applied and JIT manufacturing and planning techniques are used in a particular manufacturing system, the more that manufacturing system is edged along the continuum illustrated in Figure 16.3, i.e. towards becoming a simplified and, in the extreme case, repetitive manufacturing system. We are thinking here particularly of the matching of market requirements with a well thought out product set, the ideas of modular product design, the use of JIT manufacturing planning techniques and all of the other elements of JIT thinking discussed in Chapters 11 and 12.

16.4 MRP, OPT and JIT in the new manufacturing environment

In Chapter 1 it was argued that modern manufacturing is faced with great challenges, particularly in terms of rapid changes in customer requirements and demands, shorter product design and life cycles, and shorter lead times for deliveries to customers.

Yamashina *et al.* (1987) argue that the JIT approach is necessary in the context of increased product variety and the need to respond rapidly to customer requirements. The argument is that in the conventional approach, the manufacturing system – by installing a buffer of finished product between itself and the market – can pursue economies of scale and not be unduly influenced by changes in the marketplace in the short term. However, since product diversity increases and customer requirements change frequently, it becomes increasingly difficult to forecast which products will be sold. Moreover, operating a *buffering* policy runs the risk of having an excess of products whose demand is falling and a shortage of products that are in high demand. Figure 16.4, based on Yamashina's original presentation, illustrates the argument.

The same argument can be made for component suppliers. The manufacturer of the finished product will not want to keep large buffer stocks of raw materials and purchased parts, for reasons of economy as well as possible obsolescence, and will tend, therefore, to put pressure on his suppliers to deliver in a JIT manner. The MRP or OPT approaches, on the other hand, provide for little interaction of this type between the customer and the manufacturer.

16.5 Core theories and key failings

This section explores the core theories of the various production management paradigms – MRP/MRP II, JIT and OPT – and highlights the key failings. By its nature, this exploration involves an attempt to summarize the important elements of the three paradigms and thus follows on from the conclusions at the end of Parts II, III and IV of this book.

16.5.1 Requirements planning (MRP/MRP II)

The MRP/MRP II paradigm highlighted the fallacy of applying order point inventory control techniques to dependent demand items. Most of manufacturing activity is concerned with producing assemblies, and a bill of materials is thus a means of exploding demand from a finished product down to the components that make up the product. MRP II also showed that hierarchical planning, with multiple levels of representational detail of the manufacturing process (i.e. MPS, MRP, PAC), is a highly effective means of coping with the complexity and variety of manufacturing systems.

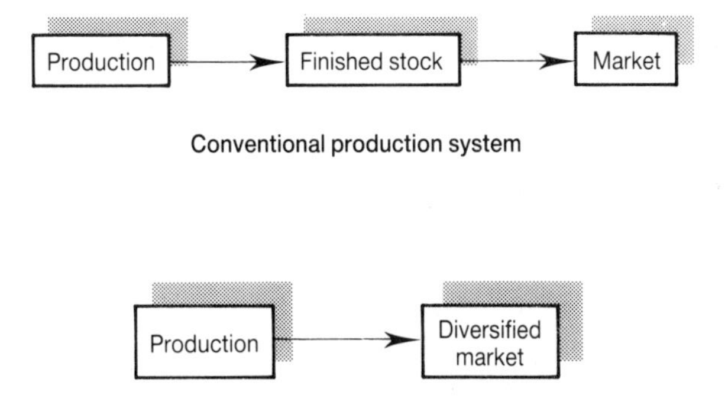

Figure 16.4 Production systems and their markets.

Another important lesson from the MRP approach is that through a computer and a manufacturing database, the work of people in many different manufacturing functions can be better coordinated, and volumes of common information can be shared. Moreover, a computer facilitates manufacturing planning to a level of detail that was never realizable before.

The MRP paradigm also stresses the role of the planner. Education must be provided to the planner. Responsibility for scheduling decisions must stay with the planner. CRP, RCCP and bottom-up replanning were designed as tools to keep the planners in charge of the planning process.

MRP/MRP II has to be seen as a reasonably successful venture, but there are qualifications to that success. The most significant is the fact that MRP did not attempt to address the design of the manufacturing process. This might not be seen to be such a major omission, but the lack of attention to the design of the manufacturing process itself leads to a situation where activities take place unnoticed (which are counterproductive to good manufacturing practice and hence to manufacturing system performance). For example, the BOM concept tended to encourage the development of many process stages, each with buffers separating them from the next stage. Now, there is nothing in MRP technology that requires this to be the case – it is just that MRP structures seem to guide users in this direction. In more recent times, the emergence of JIT has focused attention on the importance of looking at the basics of manufacturing engineering.

MRP, particularly in the development towards MRP II, has perhaps sought sophistication but achieved complexity instead. JIT has restored the pursuit of refined simplicity to its rightful position as a most worthy activity.

The MRP approach also has other faults. The idea of leaving the capacity management to the user has never worked very well. Perhaps it is not so much that the approach is wrong, but rather that the user is not in a position to take advantage of it. Capacity requirements planning, in many

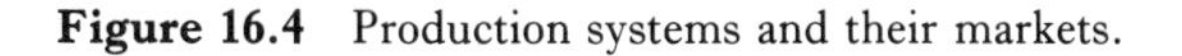

cases, overwhelms us with detail. Only now are we beginning to learn the benefits of simple aggregate resource planning (i.e. rough cut capacity planning), tools with a *what if* capability and a good user interface. The master planning level of MRP II was never really adequate on this account. Even today, there is much more to be incorporated in terms of master planning support.

Perhaps another of the faults of MRP II is that it has grown too large. It has tried to address too many problems in too many domains with the same basic approach. For example, at the shop floor level, the production activity control capability in the MRP II system is rapidly becoming redundant because of advances in CIM technology. CIM research and development has focused, to a great extent, on sophisticated control and integration systems for the shop floor. The shop floor control module of MRP II is not a viable alternative for complex CIM environments. The BOM concept may have had too much influence on the design of shop floor routings, and the price is the lack of veracity in representing manufacturing process routings. Manufacturing activities form networks, not hierarchies. Moreover, lots do need to be split.

There are other faults that can be laid at MRP's door. Some of these can be disclaimed. For example, although it is a fact that lead times cannot be predicted, this does not necessarily imply that average lead times cannot be used for planning purposes. In a sense, the use of planning lead times reflects the hierarchical nature of planning in MRP II. Actual lead times are a different matter. As long as the planning lead time is consistent with the average actual lead time, then the planning system will work reasonably well. This consistency can be maintained through the application of the rough cut capacity *what if* analysis (RCCP). The mistake, perhaps, is to try to drive production activity control with planned lead times.

The MRP approach to lot sizing is frequently criticized. In fairness, it seems that the MRP community always favoured simplicity in lot sizing or, indeed, no real lot sizing, by matching planned orders exactly with net requirements.

16.5.2 Just in time (JIT)

As stated earlier, JIT focused attention on the pursuit of manufacturing engineering excellence. Its essence is to challenge us continuously to achieve excellence, by posing ideals such as zero set-up, zero defects, zero inventories and zero lead time. JIT educated us to the fact that slack in a manufacturing system is bad because it allows mediocre behaviour and performance to pass undetected. JIT offers a philosophy of long term commitment to incremental process refinement. In many ways, the current interest in Kanban as a production control technique is amusing. Kanban is only a very small manifestation of what is a much deeper agenda.

JIT clearly showed us the dominance of engineering the manufacturing process, over planning the production. What JIT does is attempt to convert into one large system the collection of operators, equipment, raw material, etc. that make up a manufacturing organization – with the result that the activities of all the sub-systems are synchronized. With a well designed system, control becomes a less difficult problem. In repetitive manufacturing environments, control can ultimately be exercised with a manual Kanban system.

JIT also taught us about value of mixing products on the same manufacturing line, without using batches. JIT located responsibility and ownership for the manufacturing process squarely where it belongs, i.e. with the manufacturing operator. JIT also teaches us the benefits of developing flexible resources. It also shows the value of product focus and of grouping into product families to facilitate flow based production systems.

JIT, too, has its failings, though they are not yet as apparent perhaps as those of MRP II. There is a limit to the extent that JIT can be usefully pursued in many industries. The major JIT successes were in repetitive manufacturing situations. If the manufacturing system is discontinuous, in that demand is impossible to predict accurately and product variety cannot be easily constrained, then developing a JIT solution will be all but impossible. Moreover, it is not possible for all manufacturers to attain a position wherein their suppliers are both local and captive, since this phenomenon is very much a feature of the structure and state of Japanese industry.

16.5.3 Optimized production technology (OPT)

OPT teaches us that finite scheduling is a practical technology for scheduling manufacturing systems. It also illuminated the potential of recognizing bottlenecks, and of discriminating between bottlenecks and non-bottlenecks in attempts to manage the operations of the shop floor. OPT has also presented us with some very useful insights into the cost implications of scheduling decisions on the shop floor.

OPT is a proprietary technology. This, combined with the relative absence of documented case studies, naturally gives rise to some suspicion. When a technology is perceived to succeed, its success may, in fact, be due to factors other than the core technology itself, for example, the consultancy work that is tied to an OPT installation. OPT is expensive. Nonetheless, both of these latter criticisms could be applied to MRP II. However, MRP is an open book. OPT, on the other hand – in spite of our efforts – is, to a large extent, an untold story.

One other apparent weakness of the OPT paradigm stems from the fact that it does not provide as strong a sense of hierarchical planning as, say, the MRP II paradigm. OPT also seems to emphasize a technical solution to what is really a very complex organizational, as well as technical, problem.

16.6 PMS implementation issues

There can be no illusions about the difficulty of designing and installing CIM systems and, although the need for integrated manufacturing systems is accepted, there have been few such systems installed in practice.

White (1987) is of the view that 'there appear to be many reasons for failing to design and implement integrated systems'. White argues that the design and implementation of such systems requires a holistic approach, and that CIM cannot be achieved by a design process which is essentially Taylorist in nature. In this context, there are some lessons to be learned from the experiences of those who successfully implemented JIT.

A large part of the success of the JIT approach is that it involves all of those concerned in the solution to problems. For example, in the discussion on total quality control in Chapter 12, the scientific or Taylorist approach was contrasted with the JIT approach. The JIT environment was discussed in terms of multiskilled operators, trained to carry out various tasks, and reference was made to the *mutual relief* system and the role of operators and supervisors in controlling their own environment. Clearly JIT is people centred.

In Part II we quoted St. John (1984) who regretted the fact that so much effort was devoted to the lot sizing *problem*. We believe that the emphasis on *solving* this problem is clear evidence of a *technical* and reductionist approach to production management systems.

It could also be argued that OPT is in the Taylorist tradition. OPT claims to generate an optimum schedule through its proprietary algorithm and requires that the schedule be followed in every detail. Supervisors must not, in any way, interfere with it. There is no *participation* or *learning* in this approach.

There is a framework which allows us to distinguish between the traditional Taylorist approach and the approach that JIT embodies. Gault (1984) articulated this distinction when discussing the nature of Operations Research (OR) and distinguished clearly between **Technical OR** and **Socio-Technical OR**.

JIT represents, in some sense, a *socio-technical* approach to PMS and, indeed, manufacturing systems design and operation in general. The JIT approach to quality involves continuous improvement towards *zero defects*, with small groups actively seeking constant improvements on a broad range of issues, using the available know-how within the group. The emphasis on training and retraining of operators, on continuous improvement of the manufacturing process and on learning from past mistakes and failures to ensure that mistakes are not repeated, is further evidence of the approach.

The relative failure of many PMS installations can be explained, at least partially, in terms of the lack of a true socio-technical approach to the design and installation of these systems. Furthermore, many of the reasons normally advanced for disappointing results from PMS are evidence of an overemphasis on the technical aspects of PMS and a failure to give due regard to the social sub-system within which the technical sub-system has to function.

The socio-technical design approach argues that the autonomy of individuals, work groups, their work roles and the social structure within which they find themselves, are components of the organization design and structure which should be addressed, while the technical sub-systems are under development. This approach argues that the design of the social and technical sub-systems must be such as to achieve a *best fit* between the two.

We argue strongly that the design and installation of a PMS system within CIM is not a purely technical problem. We completely agree with Latham (1981) when he appeals to production and inventory management professionals to learn 'additional skills, skills in dealing with the human aspects of systems'. In a CIM environment with a relatively small, highly trained workforce, the need for this socio-technical approach is even more critical. PMS practitioners can learn much from the proponents of socio-technical design and the interested reader is referred to Cherns (1977), who lists the essential principles of socio-technical design, Trist (1982), who gives an overview and historical background to the background of the approach and Pava (1983), who discusses the application of this approach to office system design.

16.7 A recipe for the present and an image of the future

We are still faced with three *competing* technologies. How do we approach the question of choosing which strategy is appropriate for a manufacturing firm? There are some pointers in the discussion of the fundamentals of the three approaches presented earlier. If the manufacturing process is repetitive, then full blown JIT is applicable. If not, then although the JIT philosophy may yield useful process improvements, the removal of all manufacturing buffers will never be fully realizable. Thus, the extent to which one pursues JIT is determined by the nature of the products one is producing and the markets into which one is selling. There is a clear need to define the firm's manufacturing strategy in the context of the complexity of its product and its market situation.

For companies operating in a non-repetitive manufacturing environment, there is the question of what production planning strategy one should choose. Clearly, the planning process must be hierarchical. Otherwise there is simply too much detail. The question of how one planning layer connects to another is very much open. More than likely, heuristic and decision support approaches will be used at each layer. The choice between these approaches can be made, bearing in mind a number of factors. If a process is unpredictable, then finite scheduling heuristics are of less value, and it is probably best to leave the planning process primarily to the planner and provide *what if* support tools. This is also the case where the process is very much people intensive. If, on the other hand, a process is predictable in its behaviour and is machine dominated, then a heuristic finite scheduling algorithm is more likely the better way to go.

Now, of course, factories are not uniform and there may, even within the four walls of one factory, exist the need to adopt different approaches, working in coordination. We will attempt to sketch, in loose terms, the attributes of what might be called a hybrid architecture that accommodates the needs of future manufacturing systems.

It can be seen that MRP II will no longer exist in the form that is known today. Nor, indeed, will OPT. There will be a manufacturing information system that will provide database support to manufacturing decision making. Scheduling routines, including infinite scheduling with decision support or heuristic finite scheduling algorithms, will be available as policy parameters within the system, to be set by the user as he/she sees fit. Because many suppliers will neither be local nor captive, then MRP style batch scheduling has a clear role to play in acquiring purchased parts. A small percentage of suppliers will be connected to the manufacturing process by Kanban style control procedures.

On the manufacturing side, as we have seen, MRP will operate in non-predictable, human intensive, manufacturing processes, whereas finite scheduling algorithms will be used in more predictable – and machine intensive – parts of the process. If sufficient refinement is achieved, then Kanban will be used for certain parts of the process.

Sitting on top of all this will be a hierarchical master planning module. This module is unlikely to use an optimizing procedure but will, instead, be primarily a decision support tool with a library of heuristics available and, quite likely, using knowledge based support to help the master planner approach the various tradeoffs that must be made. On the shop floor level, PAC will evolve more and more towards a communications and integration tool.

We will try now to sketch the architecture for such a hybrid PMS system and present this outline architecture in terms of a hierarchy of PMS modules as illustrated in Figure 16.5.

Figure 16.5 shows PMS in terms of three sets of issues – strategic issues, tactical issues and operational issues. Let us look at each of these in turn.

16.7.1 Strategic issues

Strategic production management system issues relate to:

- the determination of the products to be manufactured,
- the matching of products to markets and customers' expectations, and
- the design of the manufacturing system,

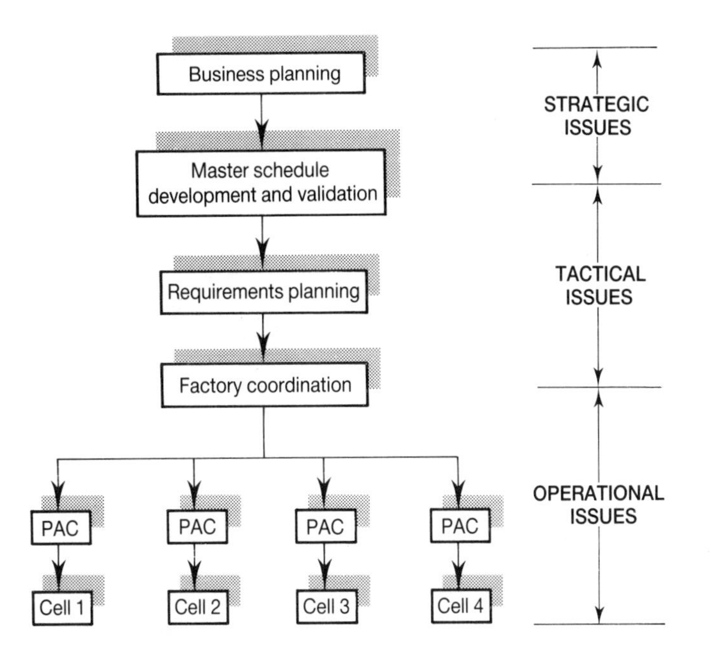

Figure 16.5 An architecture for future hybrid PMS.

to ensure short production lead times and sufficient flexibility to facilitate the production of the required variety and mix of products for the market.

It is clear that JIT thinking has a lot to offer here. It supports the strategic layer by concentrating on matching the product to the market, designing the product to facilitate manufacturing and using group technology concepts to design and define families of products, assemblies and components, to facilitate flow based manufacturing.

Ultimately, one of the outputs from the business planning activity is a master schedule to authorize production over a medium term horizon. The objective is to verify the proposed master schedule and produce an authorized master production schedule that is likely to be realizable.

If the manufacturing environment – in the broadest sense of products, markets, etc. – is repetitive, then the development and verification of the master production schedule is greatly simplified. In such cases, JIT planning techniques (as discussed in Chapter 12) are clearly appropriate.

However, it is unlikely this will be the case for the majority of manufacturing firms. What is required for non-repetitive manufacturing is a set of tools to facilitate verification from a capacity point of view, using what-if techniques embedded in a tool with a sophisticated user interface.

16.7.2 Tactical issues

Tactical production management system issues relate to the generation of detailed plans to meet the demands imposed by the master production schedule. It essentially involves the breakdown of the products in the master production schedule into their assemblies, sub-assemblies and components, and the creation of a time-phased plan of requirements, which is realistic in terms of capacity and material available.

Just in time manufacturing techniques support the requirements planning process by developing a manufacturing system which facilitates flow based manufacturing, incorporates simplicity and excellence at all stages of the manufacturing process, and develops close relationships with external suppliers. This results in a manufacturing system whose performance is highly predictable and which thus facilitates good requirements planning. In the ultimate case, this is a repetitive manufacturing system in which tactical planning can easily be achieved using production smoothing, and monthly and daily adaptation, as described in Chapter 12.

MRP supports the requirements planning process by planning the availability of material, and of manufactured and assembled parts, to meet customer requirements. MRP functions well in situations where the variety and complexity of products is such that the JIT planning techniques are inappropriate.

MRP's use of planned lead times is appropriate in the context of such relatively complex manufacturing systems. The degree to which these lead times are accurate depends on the extent to which the manufacturing system has been simplified and streamlined. The important point is that planned lead times are useful at a tactical planning level, but it should be understood that the output from this process is simply a plan, which should be interpreted as a guide for operational control purposes and not as a detailed schedule. Operation scheduling is a separate issue.

16.7.3 Operational issues

Figure 16.5 shows that we consider the factory of the future to be composed of a series of mini-focused factories or product based manufacturing cells. We believe that the degree to which such cells are *autonomous* and to which the factory can be so decomposed, will depend primarily on the product/process and market situation and the degree to which just in time thinking can be successfully applied.

Operational production management system issues essentially involve taking the output from the tactical planning phase, e.g. the planned orders from an MRP system, and managing the manufacturing system in quasi real-time to meet these requirements. Our view is that, in the future, it will be necessary to have a production activity control system for each cell and a higher level controller to coordinate the activities of the various manufacturing cells.

Let us look initially at the individual cells. We have already outlined production activity control in Chapter 4 (see COSIMA 1987). In this context, PAC is considered to be composed of three building blocks, namely, a scheduler, a dispatcher and a monitor. In our view this incorporates a very important principle that decision making be passed down to the point where most knowledge/information is available.

Let us now look at each of the building blocks in turn:

- **The scheduler** In general terms, the scheduler takes the list of required orders produced by the tactical planning system and develops schedules for each individual work centre based on the known manufacturing process routings and expected available capacity at the work centres.
- **The dispatcher** The dispatcher is, in one sense, a real time scheduler which assigns jobs to work centres based on real time information, the present status of the shop floor and on the priorities set by the scheduler.
- **The monitor** Relevant data is captured as the manufacturing process proceeds and, through the monitor block, information is fed back to the scheduler and dispatcher. In particular, this information is used by the dispatching function to facilitate real-time scheduling.

In our view, finite scheduling is appropriate at the PAC level. In particular, we believe that OPT type scheduling will find application in the scheduler building block of future PAC systems. However, we do not believe that the OPT approach, in terms of blindly following machine developed schedules, will find wide acceptance. Our concept is that one must have real-time scheduling or dispatching to allow supervisors and operators to control their environment in terms of their up-to-date knowledge of the manufacturing system.

We are not necessarily advocating that these three building blocks be software modules, although this might well be the case in many instances. One could imagine the work cell supervisor, or indeed the work cell team, dispatching work through the cell, perhaps using decision support tools.

The higher level controller which coordinates the activities of the various cells will, in our view, have a similar structure to the PAC system described above. In a sense, we see it as an upward recursion of the PAC module in that it will consist of a scheduler – perhaps using finite scheduling techniques – to schedule the planned orders, a monitor to provide feedback on cell activities and a dispatcher to manage inter-cell activity.

Just in time, in *forcing* the manufacturing system towards simplification, tries to create the environment where the problems of scheduling and dispatching are greatly reduced. This makes the problems of PAC relatively simple, so much so, that in the extreme case a manual system, namely Kanban, may be used.

We have argued already that one fault of MRP II is that it has grown too large. We believe that factory coordination, production activity control and master production schedule development will migrate to become distinct modules which will be integrated along the same lines we are proposing in the PMS architecture under discussion here.

If effort is expended on the planning procedure, then the PAC task will consequently be easier. By the planning procedure, we mean the full gamut of manufacturing planning – including product, manufacturing system and process design, etc. – as evidenced in Just in Time systems. We believe that effort expended in planning the manufacturing system pays for itself in terms of ease of control and, consequently, simplified PAC systems.

Thus, the future of PMS is probably hybrid. Whether this will occur remains to be seen, since the architectural design problems are enormous. Our architecture is simply a sketch. Future architectures may all be customized and very expensive.

16.8 Conclusions

In conclusion, our architecture is based on the following interpretations of the future of PMS. We will learn, with the coming of maturity, that JIT is a matter of degree, in that it will not be possible to achieve the ideal of just in time manufacturing across all industries. We also believe that the MRP-OPT controversy will be resolved by co-residency within a hybrid system and, finally, that both PAC and MPS will leave the MRP II domain and become separate autonomous systems in the CIM environment.

Over time, there is no doubt that a complete PMS architecture will evolve. In the meantime we believe there are rich insights – fundamental to good production management practice – to be gleaned from a study of the JIT, OPT, MRP and MRP II approaches to production management. We have tried to emphasize these insights throughout this book and we list them here for completeness (Shivnan *et al.* (1987)).

- In designing PMS systems, it is important to adopt a design methodology which gives due consideration to the social as well as the technical sub-systems of PMS.
- Product design, manufacturing system design and layout are important in the creation of an environment which will facilitate sound production and inventory management. As Burbidge (1985) points out '. . . complex production control systems do not, and probably never can, work effectively'. With intelligent product design for manufacture and assembly, it is possible, in many cases, to move in the direction of relatively simple flow based manufacturing systems. In many ways, the complexity of the production control system is dictated by the product and manufacturing system design.

- *Flow based* production systems will help shorten the production lead time and reduce inventories. Flow based manufacturing layouts are important in an era of greatly reduced product life cycles and customer expectations of rapid response to his/her demands.
- Systems designers should try to move towards decentralized planning and control which will allow much more flexibility and adaptability to changes. Ultimately, the factory should be a group of flexible manufacturing and flexible assembly cells, where each cell has been designed using group technology thinking to deal with a family of *products*. The term *product* is used rather loosely here to imply a group of identifiable assemblies or sub-assemblies of components. In so far as possible, production planning and, in particular, production control decisions should be passed down to the level which has access to pertinent data and knowledge.
- Production controllers should concentrate on balancing the flow of products through the plant, rather than attempting to balance plant capacity.
- Lead times and schedules are intimately related. Lead times are a consequence of a particular schedule and cannot be used *a priori* to generate accurate operation schedules. MRP, and its use of planned lead times, is appropriate as a tactical planning procedure.
- If effort is expended on the planning procedure, then the production activity control task will consequently be easier. By the planning procedure, we mean the full gamut of manufacturing system design and planning.
- Set-up times should be reduced as much as possible and the belief that the given set-up time is immutable and a *constant* should be discarded.
- There are many types of batches or lots. In particular, there are production lots and transfer lots. In situations where large production lots are necessary, because of limitations of the manufacturing process, the use of smaller transfer batches which facilitate flow production should be considered.
- Inventories should be reduced as far as possible and all safety and buffer stocks, which only serve to disguise problems, should be gradually reduced.
- The implementation of the above thinking requires enthusiastic support and commitment from all levels in the organization.
- Although the Kanban system is really only applicable to repetitive manufacturing situations, the JIT philosophy and manufacturing system design techniques can be applied very widely.

References

Burbidge, J.L. 1985. 'Automated production control with a simulation capability', in *Modelling Production Management Systems*, edited by P. Falster and R.B. Mazumder. Amsterdam: North-Holland.

Cherns, A.B. 1977. 'Can organization science help design organizations', *Organisational Dynamics*, Spring, 44–64.

COSIMA Project Team. 1987. 'Development towards an application generator for production activity control', in *ESPRIT 87 Achievements and Impact, Part II*, edited by The Commission of the European Communities, Directorate-General Telecommunications, Information Industries and Innovation, Amsterdam: North-Holland, 1648–1661.

Gault, R. 1984. 'OR as education', *European Journal of Operational Research*, **16**, 293–307.

Latham, D. 1981. 'Are you among MRP's walking wounded?' *Production and Inventory Management*, **23**(3), 33–41.

Pava, C. 1983. *Managing New Office Technology: An Organizational Strategy*, New York: The Free Press.

Shivnan, J., Joyce, R. and Browne, J. 1987. 'Production and inventory management techniques – a systems perspective', in *Modern Production Management Systems*, edited by A. Kusiak. Amsterdam: North-Holland, 347–362.

St.John, R. 1984. 'The evils of lot sizing in MRP', *Production and Inventory Management* **25**(4).

Trist, E.L. 1982. 'The sociotechnical perspective', in *Perspectives on Organizational Design and Behaviour*, edited by A.H. Van de Ven and W.F. Joyce. New York: John Wiley and Sons.

White, J.A. 1987. 'Integrated manufacturing systems: a material handling perspective', in *Proceedings of the 4th European Conference on Automated Manufacturing*, edited by B.B. Hundy. UK: IFS Publications, 45–56.

Yamashina, H., Okumara, K. and Matsumoto, K. 1987. 'General manufacturing strategy: the Japanese view' in *Proceedings of the 4th European Conference on Automated Manufacturing*, edited by B.B. Hundy. UK: IFS Publications, 33–44.

Bibliography

Ang, A.S. and Tang, W.H. 1975. *Probability Concepts in Engineering Planning and Design. Volume 1: Basic Principles*. New York: John Wiley and Sons.

APICS. 1983. *Training Aid – Materials Requirements Planning*, APICS.

APICS. 1984. 'JIT and MRP II: partners in manufacturing strategy', in *Report on 27th Annual APICS Conference Modern Materials Handling*, APICS, December, 58–60.

Barber, K.R. and Hollier, R. 1986. 'The effects of computer aided production control systems on defined company types', *International Journal of Production Research*, **24**(2), 311–323.

Berry, W.D. and Thompson, A.R. 1979. 'A broader view of group technology', *Computers and Industrial Engineering*, **3**, 289–312.

Bertrand, J.W.M. and Worthmann, J.C. 1982. *Production Control and Information Systems for Component Manufacturing Shops*. Amsterdam: Elsevier.

Boehlert, B. and Trybula, W. 1984. 'Successful factory automation', *IEEE Transactions on Components, Hybrids and Manufacturing Technology*, **CHMT-7** (3), 218–224.

Boothroyd, H. 1978. *Articulate Intervention: The Interface of Science and Administration*. London: Taylor and Francis.

Browne, J., Boon, J. and Davies, B. 1981. 'Job shop control', *International Journal of Production Research*, **19**(5), 633–643.

Browne, J., Chan, W. and Rathmill, K. 1985. 'An integrated FMS design procedure', *Annals of Operations Research*, **3**, 207–237.

Burstein, M.C. and Jelinek, M. 1982. 'Production management needs and information systems assumptions: a contrast', *International Journal of Operations and Production Management*, **2**(3), 37–47.

Chevalier, P.W. 1983. '*Group Technology: The Connecting Link to Integration of CAD and CAM*', Dearborn Michigan, USA: Society of Manufacturing Engineers, Technical paper MS83–506.

Corke, D. 1985. *A Guide to CAPM*. London: The Institution of Production Engineers.

Desai, D.T. 1982. 'How one firm put a group technology parts classification system into operation', *Industrial Engineering*, November, 78–86.

Fox, R.E. 1983. 'OPT. An answer for America. Part IV. Leapfrogging the Japanese', *Inventories and Production Magazine*, March–April.

Gelders, L.F. and VanWassenhove, L.N. 1985. 'Capacity planning in MRP, JIT and OPT: a critique', in *Production Economics – Trends and Issues, Proceedings of the 3rd International Working Seminar on Production Economics*, edited by R.W. Grubbstrom. Austria.

Hoyt, J. 1983. 'Determining dynamic lead times for manufactured parts in a job shop', in *Computers in Manufacturing Execution and Control Systems*. New Jersey, USA: Auerbach Publishers.

Ivanov, E.K. 1968. *Group Production, Organization and Technology*. UK: Business Publications.

Jaikumar, R. 1986. 'Postindustrial manufacturing', *Harvard Business Review*, November–December, 69–76.

Klein, L. 1986. 'Manufacturing planning and control: who's responsible?', *CIM Technology*, **8**(3), 13–15.

Kochhar, A. 1979. *Development of Computer Based Production Systems*. London: Edward Arnold.

Krupp, J.A.G. 1987. 'Rough cut bottleneck master scheduling', *P&IM Review*, May, 34–36.

Lockyer, K.G. 1984. 'Production management – the art of the possible', *International Journal of Operations and Production Management*, **4**(4), 28–36.

Nagarkar, C.V. and Fogg, B. 1979. 'Application of group technology to manufacture of sheet metal components', *Annals of the CIRP*, **28**(1), 407–411.

New, C. 1985. 'Manufacturing in the 1980s', in *Manufacturing Systems Context, Applications and Techniques*, edited by V. Bignell, M. Donnere, J.Hughes, C. Pym and S. Stone. UK: Basil Blackwell in association with The Open University.

Plossl, G. and Wight, O. 1967. *Production and Inventory Control Principles and Techniques*. New Jersey, USA: Prentice–Hall.

Ranky, P. 1986. *Computer Integrated Manufacturing: An Introduction with Case Studies*. London: Prentice–Hall International.

Regan, J. 1983. *Design of a Microcomputer Based Net Change MRP System*, M.Eng.Sc. thesis, University College Galway, Ireland.

Sata, T. 1984. 'A view of the highly automated factory of the future', *Robotics and Computer Integrated Manufacture*, **1**(2), 153–159.

Taylor, J.C. 1976. *A Report of Preliminary Findings from the 1976 Work Organization Pilot Study*. UCLA, USA: Centre for the Quality of Working Life.

Voss, C.A. 1987. *Just In Time Manufacture*. UK: IFS Publications.

Wild, R. 1983. *Production and Operations Management: Principles and Techniques*, 3rd Edition. UK: Holt Rinehart and Winston.

Yoshikawa, M., Rathmill, K. and Hatvany, J. 1981. *Computer Aided Manufacturing, An International Comparison*. Washington, DC: The National Academy Press.

Glossary of acronyms

AGV	Automatic Guided Vehicle
AI	Artificial Intelligence
AMHS	Automated Materials Handling Systems
AMHSS	Automated Materials Handling and Storage Systems
APICS	American Production and Inventory Control Society
AQL	Acceptable Quality Level
AS/RS	Automatic Storage and Retrieval System
BOM	Bill of Materials
BOMP	Bill of Materials Processing
CAD	Computer Aided Design
CAE	Computed Aided Engineering
CAM	Computer Aided Manufacture
CAM-I	Computer Aided Manufacturing-International, Inc.
CAPM	Computer Aided Production Management
CAPP	Computer Aided Process Planning
CAT	Computer Aided Test
CFA	Company Flow Analysis
CIB	Computer Integrated Business
CIM	Computer Integrated Manufacturing
CNC	Computer Numerical Control
CRP	Capacity Requirements Planning
DIP	Dual In-line Package (electronic component)
DNC	Direct (or Distributed) Numerical Control
DP	Data Processing
EBQ	Economic Batch Quantity
EDI	Electronic Data Interchange
EOQ	Economic Order Quantity
ESPRIT	European Strategic Programme for Research and Development in Information Technology
FFA	Factory Flow Analysis
FMC	Flexible Manufacturing Cell
FMS	Flexible Manufacturing System
GA	Group Analysis
4GLs	Fourth Generation Languages
GT	Group Technology

JIT Just In Time
LA Line Analysis
LTPD Lot Tolerance Percent Defective
MAP Manufacturing Automation Protocol
MIS Management Information Systems
MPS Master Production Schedule
MRP Materials Requirements Planning
MRP II Manufacturing Resource Planning
NC Numerical Control
OPT Optimized Production Technology
OR Operations Research
PAC Production Activity Control
PBC Period Batch Control
PFA Product Flow Analysis
PMS Production Management System
POQ Periodic Order Quantity
RCCP Rough Cut Capacity Planning
SMED Single Minute Exchange of Dies
TA Tooling Analysis
TQC Total Quality Control
VCD Variable Centre Distance (electronic component)
VLSI Very Large Scale Integration
WIP Work in Progress

Index